Sustainable Civil Infrastructures

About this Series

Sustainable Infrastructure impacts our well-being and day-to-day lives. The infrastructures we are building today will shape our lives tomorrow. The complex and diverse nature of the impacts due to weather extremes on transportation and civil infrastructures can be seen in our roadways, bridges, and buildings. Extreme summer temperatures, droughts, flash floods, and rising numbers of freeze-thaw cycles pose challenges for civil infrastructure and can endanger public safety. We constantly hear how civil infrastructures need constant attention, preservation, and upgrading. Such improvements and developments would obviously benefit from our desired book series that provide sustainable engineering materials and designs. The economic impact is huge and much research has been conducted worldwide. The future holds many opportunities, not only for researchers in a given country, but also for the worldwide field engineers who apply and implement these technologies. We believe that no approach can succeed if it does not unite the efforts of various engineering disciplines from all over the world under one umbrella to offer a beacon of modern solutions to the global infrastructure. Experts from the various engineering disciplines around the globe will participate in this series, including: Geotechnical, Geological, Geoscience, Petroleum, Structural, Transportation, Bridge, Infrastructure, Energy, Architectural, Chemical and Materials, and other related Engineering disciplines.

More information about this series at http://www.springer.com/series/15140

WJvdM Steyn · Irina Holleran
BooHyun Nam
Editors

Pavement Materials and Associated Geotechnical Aspects of Civil Infrastructures

Proceedings of the 5th GeoChina International Conference 2018 – Civil Infrastructures Confronting Severe Weathers and Climate Changes: From Failure to Sustainability, held on July 23 to 25, 2018 in HangZhou, China

Springer

Editors
WJvdM Steyn
Department of Civil Engineering
University of Pretoria
Pretoria, Gauteng
South Africa

BooHyun Nam
Department of Civil, Environmental, and Construction Engineering
University of Central Florida
Orlando, FL, USA

Irina Holleran
Department of Civil and Environmental Engineering
Auckland University
Auckland, New Zealand

ISSN 2366-3405 ISSN 2366-3413 (electronic)
Sustainable Civil Infrastructures
ISBN 978-3-319-95758-6 ISBN 978-3-319-95759-3 (eBook)
https://doi.org/10.1007/978-3-319-95759-3

Library of Congress Control Number: 2018948645

This Springer imprint is published by the registered company Springer Nature Switzerland AG
The registered company address is: Gewerbestrasse 11, 6330 Cham, Switzerland

Contents

Introduction

This volume contains nine papers that were accepted and presented at the GeoChina 2018 International Conference on Sustainable Civil Infrastructures: Innovative Technologies and Materials, to be held in HangZhou, China, July 23 to 25, 2018. It contains research data, discussions, and conclusions focusing on a number of pavement materials and related geotechnical aspects of infrastructure. Topics include issues related to civil infrastructure such as moisture sensitivity, soil, slope, and rock stability and sustainable pavement construction, monitoring, and maintenance. This information should lead to more resilient infrastructure design, maintenance, and management. Various types of research were used in the various studies, including field measurements, numerical analyses, and laboratory measurements. It is anticipated that this volume will support decisions regarding the optimal management and maintenance of civil infrastructures to support a more resilient environment for infrastructure users.

Introduction

Development and Application of Water-Saving and Moisture-Retaining Membrane Made from Controllable High Polymer Materials for Concrete Curing

Jia-liang Yao[1,2(✉)], Zhen-quan Wang[1], and Dong-han Tang[3]

[1] School of Traffic and Transportation Engineering, Changsha University of Science and Technology, Changsha 410114, Hunan, China
yao26402@126.com, 1970229819@qq.com

[2] Key Laboratory of Road Structure and Material of Ministry of Transport (Changsha), Changsha University of Science and Technology, Changsha 410114, Hunan, China

[3] Hunan Science and Technology Research and Development Institute, Changsha 410004, Hunan, China
398290355@qq.com

Abstract. The conventional cement concrete pavement curing is frequently criticized for large water consumption, high labor intensity, high cost, non-uniformed curing quality and other problems. With an attempt to solve these problems, the research of this paper, supported by some concrete pavement construction projects and some concrete pillar construction projects in China, explored to develop a new type of curing membrane from the controllable high polymer materials for the concrete curing to achieve the water saving and moisture retaining. The curing membrane for water-saving and moisture-retaining was developed with a new type of controllable high polymer as the interlayer material, with a specialized plastic membrane as the carrier, and with the concrete hydration principle as the theoretical background. In the research of this paper, a series of tests were done to measure temperature and humidity of the concrete cured by different methods. Furthermore, strength, carbonization depth and cracking of the concrete slabs under different types of curing were also tested. The results indicated that (1) the use of the new-type water-saving and moisture-retaining membranes to cure the concrete for 28 days could retain the surface humidity of concrete up to greater than 90%; (2) the microcracks of the concrete slabs during summer construction were reduced by the combined use of a curing agent and the new-type water-saving and moisture-retaining membrane; and (3) compared with the natural curing, carbonization depth of the concrete cured by the new-type curing membrane for 90 days was reduced by 60% while the concrete strength increased by 8.6 MPa, up to 51.6 MPa.

Keywords: Cement concrete pavement · Concrete pillar · Curing membrane Construction · Water saving and moisture retaining

W. Steyn et al. (eds.), *Pavement Materials and Associated Geotechnical Aspects of Civil Infrastructures*, Sustainable Civil Infrastructures,
https://doi.org/10.1007/978-3-319-95759-3_1

1 Introduction

Portland cement should set and harden under the appropriate humidity and temperature. Otherwise, the lack of necessary moisture curing will lead to the degradation of concrete performances such as volume stability, compressive strength, and impermeability (Ramezanianpour and Malhotra 1995; Khatib and Mangat 2002; Wu and Lian 1999). In this regard, the concrete must be cured shortly after its pouring and molding. Curing is a must to prevent the concrete from abnormal shrinkage, cracks and other damage caused by exposure to the burning sun, wind, dehydration, coldness and other natural factors after the formation of concrete (Nevile 1996). Curing provides cement (more precisely, the cementitious material in cement) with the appropriate temperature and humidity for its hydration. During construction, strength of concrete is controlled on the one hand through the choice of raw materials and its mix ratio design, on the other hand through the reasonable curing measures to ensure the hydration of cement and other cementitious materials. With improper curing, even if the concrete is of high quality, it will be vulnerable to carbonation, thereby reducing the durability of concrete structures (Huang and Yang 2004; Yang et al. 2004).

The research of this paper was to develop a new type of curing membrane from the controllable high polymer materials for concrete curing, with an attempt to solve some problems in the conventional curing method such as large water consumption, high labor intensity, high cost, non-uniformed curing quality and others. A series of tests were done in some concrete pavement construction projects and some concrete pillar construction projects in China.

2 Development of New-Type Curing Membrane from High Polymer Materials

If the conventional plastic membrane is used to cover concrete for curing, the plastic membrane is adverse to moisture retention of the concrete because the excess water from the concrete construction will turn to be water vapor due to the high temperature within the cover of membrane. Meanwhile, the water vapor will increase the air pressure within the cover of membrane so as to lift the plastic membrane from the concrete surface. Therefore, the conventional plastic membrane is not conducive to water retention and is also not resistant to wind blowing so as to be damaged easily. If sacks or geotextile are used to cover concrete for curing, it is necessary to water frequently during the curing, which leads to greater water consumption and higher labor intensity. The key to solving the above-mentioned two problems is to find a curing material (1) which can make the liquid water in concrete under its cover not turn to be water vapor with rising of temperature; and (2) which can store the moisture volatilized by the concrete and release the moisture when the concrete hardening needs it so as to maintain the required humidity for the concrete curing. The development of this new-type curing material was the target of the research of this paper.

2.1 Water-Absorbing Property of New-Type Cuing Material from Controllable High Polymer

After a large number of tests, the research of this paper chose to use the controllable polymer absorbent material as the curing membrane. The property of water absorption is the first consideration for the choice of controllable polymer absorbent material. In principle, the material with higher water absorption is preferred. But the higher the water absorption, the more soft the polymer material, the more easily degraded. Material without degradation is difficult or less to absorb water; but material with fast degradation is of shorter life. To strike a balance, super absorbent resin was chosen. Super absorbent resin is a strong hydrophilic polymer compound of three-dimensional network structure and low cross-linking. Because of its special molecular structure, the super absorbent resin can store within its polymer network as much water as the 100–1000 times of its own weight (Kong and Li 2009).

Three kinds of curing material from super absorbent resin (represented by A, B, and C) were selected for water absorption tests. The results are shown in Table 1.

Table 1. Water absorption of different super absorbent materials

Absorbent material	Multiple of the initial mass after adsorbing deionized water	Multiple of the initial mass after adsorbing tap water	Time (T1) for adsorbing water up to 50% of saturation (Min)	Time (T2) for adsorbing water up to full saturation (Min)
Product A	780	402	5	20
Product B	600	300	4	16
Product C	500	254	5	14

2.2 Degradation of Different Curing Materials

To meet the environmental requirements, the water-saving and moisture-retaining curing materials should be degradable after being abandoned. The above-mentioned three kinds of curing material A, B and C were tested for their degradation rate under outdoor temperature of 33 °C and average humidity of 50%. The results indicated that for A, its degradation rate on the third day was 90%; for B, it was 45%; and for C, it was 20%.

To strike a balance between water absorption and degradation, different types of resin were compounded. It was found that the BC synthesis was desirable for being used as the interlayer material of the new-type curing membrane because its multiple of the initial mass after adsorbing deionized water can be controlled within 300–400 while it is of acceptable degradation.

2.3 Structure of Curing Membrane and Its Performances

The desirable curing membrane is expected to be of faster water absorption, better moisture retention and thermal insulation, resistance to light pressure, easy storage and transportation, and acceptable cost. Through a series of experimental comparison, a sandwich structure (see Fig. 1) was adopted for the new-type curing membrane researched in this paper.

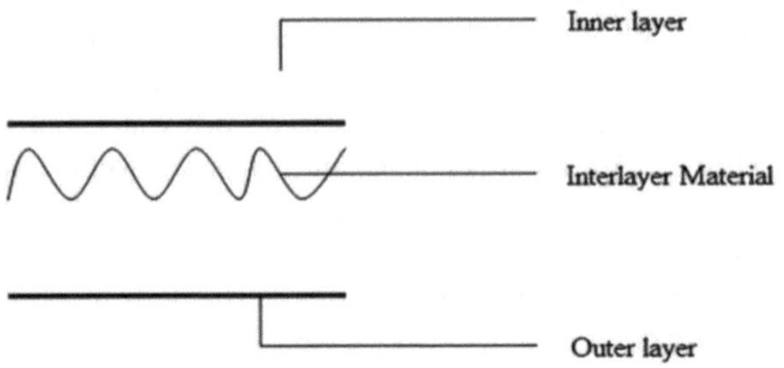

Fig. 1. Sketch for the sandwich structure of the new-type curing membrane

The curing membrane was composed of the inner, middle and outer layers. The inner and outer layers were distributed with small holes with moderate size and uniform spacing. The outer layer had a certain capability to prevent the degradation of the super absorbent material during the curing of the concrete, and also to reduce the evaporation of water. A variety of outer layers from different materials were tested for their time to keep water from evaporation and their moisture-retaining time. The test results are listed in Tables 2 and 3 respectively (1, 2, 3 and 4 representing different products of outer layer).

Table 2. Time for moisture retention of different outer layers

Outer layer	Time for moisture retention (h)
Product 1	20
Product 2	16
Product 3	38
Product 4	25

Table 3. Cumulative water evaporation rate of different outer layers

Outer layer	After 4 h. of evaporation (%)	After 8 h. of Evaporation (%)	After 12 h. of evaporation (%)
Product 1	10	30	42
Product 2	14	35	46
Product 3	8	18	27
Product 4	20	45	57

The test results show that the Product 3 is of the longest moisture retaining time and the lowest evaporation rate.

The interlayer material is a mixture of polymer absorbent material and carrier carrying the polymer absorbent material. In the research of this paper, 5 types of interlayer materials were tested for their water absorption, which was indicated by the swelling speed of the interlayer materials after being watered. The test results are shown in Table 4. It can be seen from Table 4 that the Product 2 was of the shortest time (2 min) to make the curing membrane swell after absorbing water; while the Product 5 was of the longest time (8 min) to make the curing membrane swell after absorbing water.

Table 4. Time needed for different interlayer materials to swell after absorbing water

Interlayer material	Time needed to swell (min)
Product 1	5
Product 2	2
Product 3	6
Product 4	3
Product 5	8

There are two options for the inner layer of the new-type curing membrane—agricultural film and regenerated membrane, depending on the market demand. If the curing membrane is to be used many times, combined with the less degradable absorbent material and the high-strength carrier as the interlayer material, the agricultural film is a better choice as the inner layer thanks to its better tensile strength and compressive strength. The properties of the two types of material for the inner layer are displayed in Table 5.

Table 5. Properties of two types of material for inner layer

Inner layer	Tensile strength (kg/cm^2)	Compressive strength (kg/cm^2)	Elongation (%)
Agricultural film	140–180	100–120	70–90
Regenerated membrane	120–160	80–110	40–70

Whether the curing membrane is for one-time use or repeated use, the inner layer is so relatively thin that the curing membrane can be fixed to the concrete surface without being blown open after water is poured on it. The indoor simulated wind test was done with the help of an industrial fan with an adjustable wind speed. During the test, the industrial fan was located at a horizontal distance of 50 cm from the curing membrane, facing the obverse side of the experimental specimens. The wind speed was set at the following three levels: Level 1 of 5 m/s, Level 2 of 10 m/s, and Level 3 of 20 m/s. The test results are listed in Table 6.

Table 6. Wind resistance of curing membranes with different types of inner layer

Inner layer	At wind speed of 5 m/s	At wind speed of 10 m/s	At wind speed of 20 m/s
Agricultural film	No open	No open	No open
Regenerated membrane	No open	No open	No open
Membrane of 0.02 mm thickness	Open occasionally	Open	Open

As the curing membrane can be fixed to the cured surface tightly and the edges of the inner layer and the outer layer of the membrane are sealed together, the entire curing membrane can form a relatively complete independent system so that it has a better wind resistance and thermal insulation. In other words, the concrete covered with the new-type curing membrane will heat up and cool down steadily, with less risk of temperature cracks caused by a sudden rise or fall of temperature. Furthermore, the super absorbent material (so-called "small reservoir") contained in the interlayer material is of less occurrences of water evaporation thanks to the better encapsulation of the membrane. To sum up, the new-type curing membrane is not only much better in moisture retention than sacks, straw, sawdust and others, its comprehensive curing effect is also greatly improved.

3 Mechanism and Technical Characteristics of New-Type Curing Membrane

3.1 Mechanism Behind Water Absorption and Moisture Retention of New-Type Curing Membrane

The mechanism for the new-type curing membrane to absorb water and retain moisture well is explained as follows. The polymer absorbent material adhering to the inner layer of the new-type curing membrane is of strong water absorption and moisture retention. When it absorbs water to saturation, it will swell rapidly (as shown in Fig. 2). Thanks to a strong effect of moisture retention and thermal insulation of the outer layer, a humid and warm environment is created at the surface of the concrete covered by the curing membrane. With the continued hydration of cement, the free water in the concrete disappears, and the surface humidity is lowered. At this point, the polymer absorbent material close to the concrete surface can release the absorbed moisture to the concrete surface, and the moisture may penetrate into the interior of the concrete through the capillary action so as to promote the hydration of cement. In addition, the thermal insulation effect of the outer layer will make the temperature inside the membrane rise and speed up the hydration process, increasing the hydration products at early age and making the concrete structure more compact through lasting hydration at later age. These effects have been reported by some researches. In a research (Guo and Kong 2010), three concrete slabs (80 cm × 40 cm × 10 cm) were placed in the outdoor environment, and cured respectively by three different curing methods: no curing at all as a comparison (Method A); curing with the new-type water-saving and

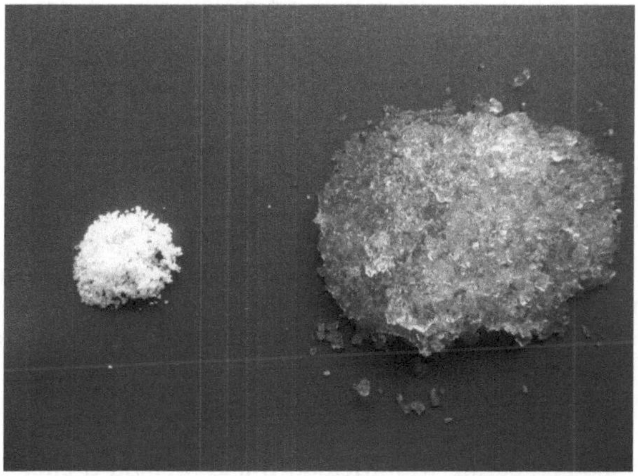

Fig. 2. Change of the appearance of the polymer absorbent material before and after absorbing water

moisture-retaining curing membrane developed in the research of this paper (Method B); and curing with another kind of curing membrane produced in Beijing after being immersed in water to full saturation (Method C). After 7d, samples taken from the concrete specimens in the area within 5 mm close to the curing membranes were tested by the mercury intrusion porosimetry (MIP). The test results about porosity of different concrete slabs cured with Method A, Method B, and Method C are listed in Table 7, and the pore structure of different concrete slabs cured with Method A, Method B, and Method C are reported in Table 8.

The pore size distribution of concrete has an important effect on the durability of concrete. A research (Wu and Lian 1999) claimed that the pore of concrete can be divided into four grades in terms of its diameter: the harmless pore of the diameter less than 20 nm, the pore of less harm with the diameter 20–50 nm, the harmful pores of the diameter 50–500 nm, and the extremely harmful pores of the diameter more than 200 nm.

Table 7 indicates that the porosity of the concrete cured with Method B and Method C was small. And Table 8 shows that the percentage of the harmful and extremely harmful pores of the concrete cured with Method B was significantly lower

Table 7. Porosity of concrete slabs cured by different methods

Curing method	Total volume of mercury pressed into concrete (ml/g)	Porosity (%)
A	0.0971	20.22
B	0.1052	20.68
C	0.0971	14.11

Notes Method A refers to no curing at all; Method B refers to curing with the new-type water-saving and moisture-retaining curing membrane developed in the research of this paper; and Method C refers to curing with another kind of curing membrane produced in Beijing after being immersed in water to full saturation

Table 8. Percentage of Pore size distribution of concrete slabs cured by different methods

Curing method	Pore of diameter <20 nm	Pore of diameter 20–50 nm	Pore of diameter 50–200 nm	Pore of diameter >200 nm
A	29.2	22.9	17.0	30.9
B	34.4	54.8	3.5	7.3
C	27.7	40.6	13.6	18.1

Notes Method A refers to no curing at all; Method B refers to curing with the new-type water-saving and moisture-retaining curing membrane developed in the research of this paper; and Method C refers to curing with another kind of curing membrane produced in Beijing after being immersed in water to full saturation

than that with Method A; curing with Method B and Method C effectively reduced the harmful and extremely harmful pores in concrete and thereby could promote strength and carbonization resistance of concrete, which would help improve the durability of concrete structures.

3.2 Technical Characteristics of New-Type Curing Membrane

The key raw material of the new-type curing membrane is a new controllable polymer absorbent material—starch grafted acrylonitrile. The research of this paper developed the curing membranes to cure the horizontal surfaces and the vertical surfaces of concrete respectively, including a series of products such as ones for curing concrete pavements exclusively, for curing cement stabilized base exclusively, for curing the vertical surfaces of concrete exclusively and other products.

The technical characteristics of the new-type curing membrane are shown in Table 9.

4 Tests for Concrete Cured with New-Type Curing Membrane

Considering the importance of temperature and humidity control in the process of concrete construction, temperature and humidity of the concrete covered by the curing membrane were measured by the research. Meanwhile, what were measured included compressive strength and carbonization depth of concrete at different age and cured by different method such as curing with the new-type curing membrane, conventional curing and natural curing. Also, cracking on the concrete surface was observed and measured.

4.1 Tests for Temperature and Humidity of Concrete Covered by Curing Membrane

Humidity within the cover of membrane can reflect the effect of concrete curing. Higher the humidity, better the curing effect. In addition, the temperature difference

Table 9. Technical characteristics of the new-type curing membrane

Curing membrane	Appearance	Breadth (m)	Carrier (g/m^2)	Resin (g/m^2)	No seal at the edge	Aperture (mm)	Hole spacing (cm)	Water retention rate after 3d (%)
First grade product	Gray uniform membrane	1.05 ± 0.01	50 ± 5	8–15	No	Φ2–4	(6.0 ± 0.5)χ(6.0 ± 0.5)	95
Qualified product	Gray membrane without leaking white basically	1.05 ± 0.03	50 ± 5	8–15	≤5 cm, no more than 2 places per meter	Φ2–4	(6.0 ± 0.5)χ(6.0 ± 0.5)	85

between the inside and the outside of membrane should not be large. Otherwise, the large temperature difference will lead to the deformation of concrete and then cracking at early age. The practical experiences show that the temperature difference is best less than 10 °C. In this regard, supported by the Shao-huai Expressway Project in Hunan Province, the research tested temperature and humidity of the inside and the outside of the curing membrane at the different curing period. The test results are listed in Table 8. No cracks and broken slabs occurred at the 64.1 km-long cement pavement of the Shao-hui Expressway Project during the construction.

The Shao-huai Expressway was constructed from May 13 to June 10 in 2007. The curing period lasted 28 days. The pavement was covered with the new-type curing membrane about 8 h later after the initial setting. From that time on, temperature and humidity of the inside and the outside of the curing membrane at the different curing period within a day were measured (see Fig. 3). The start of the time to measure was set with the consideration that hydration and heat release of concrete will reach the peak at this period of time, and temperature and humidity have great effect on the performances of concrete at early age. Thus the test results are favorable to choosing the proper curing membrane for the similar concrete projects in the future.

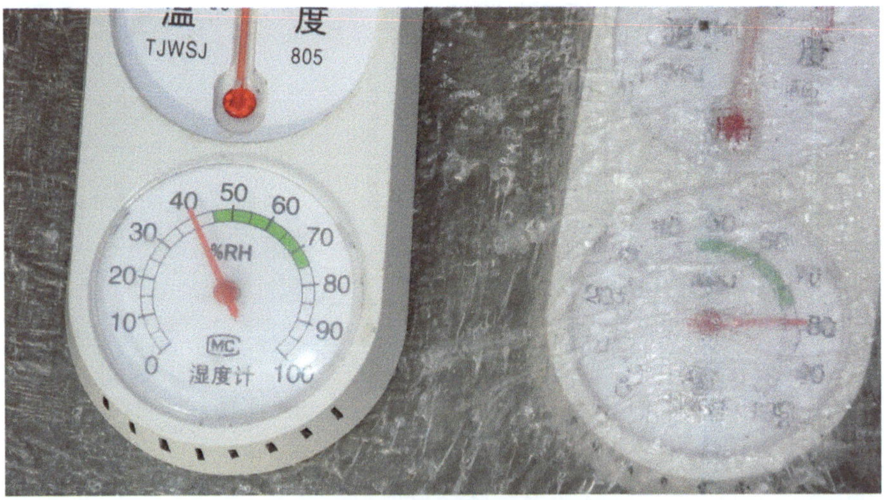

Fig. 3. Measurement of temperature and humidity of the inside and the outside of curing membrane. *Notes* The curing membrane was covered over concrete about 8 h later after the initial setting. The measuring time was from 0:10 to 17:00. The temperature during construction was 15–38 °C and the concrete ambient temperature was 18–44 °C

The measurement results (as shown in Table 10) indicated that the use of water-saving and moisture-retaining curing membrane could keep the concrete surface humidity greater than 90%. Actually, the curing membrane was of two colors—black and white. The data shown in Table 8 are about the white curing membrane. Because the white membrane could reduce the radiant heat of sunlight, at the outside ambient

Table 10. Temperature and humidity of inside and outside of curing membrane

Time to measure	Outside ambient temperature (°C)	Inside temperature (°C)	Outside ambient relative humidity (%)	Inside relative humidity
0:10	23	31.3	64	96
0:05	18	24	68	93
0:30	23	27	64	100
0:15	21	25	65	92
0:50	22	27	64	91
7:00	25	29	71	97
8:00	25	35	74	97
8:20	26.5	31.9	70	95
8:40	20	25	68	95
9:00	25	27.5	60	91
12:00	40	49.8	56	99
13:00	40	44	60	100
14:00	40	48	62	100
14:30	44	52	58	100
16:00	35	40	60	98
16:30	36	39	60	95
17:00	33	42.3	60	98

temperature of 20–44 °C, the temperature difference between the inside and the outside of the white membrane was 10 °C less than that of the black membrane. Therefore, the white curing membrane is preferable for the construction in the high temperature, while to meet the need for thermal insulation in the lower temperature, the choice of the black curing membrane is favorable.

When the ambient temperature is high, if temperature of the curing water is low, it will increase the temperature difference between the concrete surface and the interior concrete, which may cause so excessive temperature stress of the concrete that the surface cracks may occur during the concrete curing. Table 8 indicates that the relative humidity inside the membrane was stable and more than 90%, and the temperature difference between the inside and the outside of the membrane was less than 10 °C.

4.2 Tests for Splitting Strength of Cores Drilled from Concrete Pavement at Early Age

The cement concrete pavement in the above-mentioned Shao-huai Expressway Project was cured with the new-type water-saving and moisture-retaining membrane for 28 days. And then, the pavement was cored for the splitting strength test. The test results are shown in Table 11.

Table 11. Converted bending strength of core specimens drilled from concrete pavement at 28d age

Stake to drill core specimen	Converted bending strength of core 1 (MPa)	Converted bending strength of core 2 (MPa)	Converted bending strength of core 3 (MPa)	Converted bending strength on average (MPa)
Right side of K13+480	6.60	7.66	6.87	7.04
Right side of K12+705	8.04	7.12	7.23	7.46
Right side of K7+810	7.00	7.09	6.47	6.85
Left side of K11+130	7.56	7.60	7.63	7.60
Left side of K14+350	7.13	7.27	6.93	7.11
Right side of K23+575	6.87	6.87	7.33	7.02
Left side of K14+455	7.07	7.33	7.60	7.33
Left side of K16+615	7.13	7.69	6.67	7.16
Left side of K26+600	6.61	6.83	7.07	6.69
Right side of K26+000	7.25	6.99	7.71	7.32
Left side of K25+400	6.67	6.48	6.40	6.52
Left side of K30+965	7.33	7.95	8.00	7.76
Right side of K29+200	6.60	6.53	7.09	6.74
Right side of K25+600	8.33	8.40	7.60	8.11
Left side of K38+490	8.07	7.67	7.87	7.87
Left side of K46+250	6.67	6.40	7.33	6.80
Right side of K46+675	5.80	6.40	6.40	6.20
Right side of K44+500	7.07	6.60	7.27	6.98
Right side of K51+755	7.00	7.33	7.20	7.11
Right side of K52+150	6.73	7.00	7.53	7.09

(continued)

Table 11. (*continued*)

Stake to drill core specimen	Converted bending strength of core 1 (MPa)	Converted bending strength of core 2 (MPa)	Converted bending strength of core 3 (MPa)	Converted bending strength on average (MPa)
Right side of K49+300	6.40	6.20	6.67	6.42
Right side of K54+690	7.47	8.40	8.00	7.96
Right side of K52+940	6.51	6.19	5.97	6.22
Right side of K57+415	7.13	7.00	8.47	6.87

The designed bending strength of the Shao-huai Expressway was 5.5 MPa. According to the *Quality Inspection and Evaluation Standards for Highway Engineering* (JTG F80/1-JTG 2004), the average strength is calculated by

$$\bar{f} \geq \mathrm{f_d} + k\sigma \tag{1}$$

And the minimum strength is calculated by

$$f_{\min} \geq 0.85fd \tag{2}$$

It is evident that the bending strength of the above core specimens was qualified.

4.3 Tests for Cracking of Concrete Slabs Cured by New-Type Curing Membrane

In the tests for cracking, concrete slab specimens were casted following the concrete mix ratio design and the design bending strength (5.5 MPa) of another concrete project —the Qinglin Expressway Project in Guangdong Province. Three methods were taken to cure these slab specimens for comparison: no curing at all (Method A), curing with a curing agent (Method B) and curing with a curing agent plus the new-type water-saving and moisture-retaining membrane (Method C). All slab specimens cured with the three different curing methods were tested for cracking.

The test method for cracking was first proposed by Kasai Yifu (Yokoyama et al. 1994). The four sides of molds used for molding the test specimens were made of 63 mm-high section steel. And the rebar sticks (Φ10 mm × 100 mm) were installed around the mold inside so that the specimen would be restrained when the specimen was contracted. The specimen size was 600 mm × 600 mm × 63 mm. The specimens were tested within the molds. Considering the geographical environment and climate of the location of the Qinglian Expressway Project, the research of this paper used heating to replace the electric fan blowing in the Kasai Yifu's test method (see Fig. 4) to simulate the realistic climatic conditions of the concrete pavement.

Fig. 4. Test for cracking of concrete slabs

During the test, the cracking time of the specimens, the number of cracks, the length and the width of the cracks were recorded. The record of the cracking time of the specimens lasted 24 h from the time to cast concrete. Based on the 24-h record of cracking, the following three parameters are calculated:

(1) average cracking area,

$$\frac{1}{2N}\sum_{i=1}^{n} WiLi\ (\mathrm{mm}^2) \tag{3}$$

(2) number of cracks per unit area,

$$\mathrm{b} = \mathrm{N/A\ per\ m}^2 \tag{4}$$

and (3) total cracking area per unit area (m^2).

$$\mathrm{c} = \mathrm{a} \times \mathrm{b}\ (\mathrm{mm}^2/\mathrm{m}^2) \tag{5}$$

Where: Wi = maximum width of the $i\text{-}^{th}$ crack (mm); Li = length of the ith crack (mm); N = total number of cracks; and A = area of test specimen, 0.36 m^2.

The test results of the research of this paper are listed in Table 12, in which the grade of crack resistance of concrete was assessed in accordance with Kasai Yifu's method. Kasai proposed that the evaluation criteria to grade the crack resistance at early age included: (1) only very fine cracks, (2) average cracking area (a) <10 mm^2, (3) Number of cracks per unit area (b) <10 per m^2, and (4) the total cracking area per unit area (c) <100 mm^2/m^2. In accordance with the above criteria, the crack resistance of concrete was graded as five: Grade I (completely meeting the above four criteria);

Grade II (meeting three of the above four criteria); Grade III (meeting two of the above four criteria); Grade IV (meeting one of the above four criteria); and Grade V (meeting none of the above four criteria) (Yokoyama et al. 1994).

Table 12. Crack resistance of concrete cured by different methods at early age

Curing method	Room temperature (°C)	Maximum width of crack (mm)	Average cracking area (mm^2)	Number of cracks Per unit area (per m^2)	Total cracking area per unit area (mm^2/m^2)	Grade of crack resistance
A	26	0.42	6.87	55.56	381.23	IV
B	27	0.26	7.23	50.00	137.28	IV
C	26	0.09	6.47	8.33	10.21	I

Notes Method A refers to no curing; Method B refers to curing with a curing agent; and Method C refers to curing with a curing agent plus the new-type water-saving and moisture-retaining curing membrane

Table 12 indicates that (1) no curing made the concrete specimens under the test conditions of this research cracked seriously, in line with Grade IV of crack resistance, and with wider cracks; (2) for the specimens cured with spraying an emulsified wax curing agent, its crack resistance was also of Grade IV, but with less width of cracks and smaller average cracking area; and (3) for the specimens cured by spraying the emulsified wax curing agent at extremely early age, and then covering with the new-type water-saving and moisture-retaining curing membrane while the specimens were heated (i.e., a kind of compound curing), cracking did not occur 24 h later, and its crack resistance was in line with Grade I.

4.4 Tests for Strength, Carbonization Depth and Cracking of Concrete Pillars Cured by New-Type Curing Membrane

If the water-saving and moisture-retaining membrane aimed for curing the horizontal surfaces of concrete structures is used for curing the vertical surfaces, when the polymer absorbent material of the membrane is saturated with water, its weight and gravity increase so much that it cannot overcome its gravity and will drop down. To solve the problem, the research of this paper developed a specialized curing membrane YHM-L/204 for curing the vertical surfaces of concrete. After the vertical surfaces of concrete were cured with YHM-L/204 to the specified age, they were tested for strength (by the rebound method) and carbonization depth.

The concrete strength grade of the concrete pillars in the supporting Project is C30. And the pillar size is 0.65 m × 0.65 m × 3.50 m. The YHM-L/204 curing membrane was paved in accord with the specifications for construction of the curing membrane. Figure 5 displays the concrete pillar wrapped with the YHM-L/204 curing membrane.

After the concrete pillar was cured to the specified age by different curing methods (natural curing, conventional curing and curing with the YHM-L/204 curing membrane respectively), tests for concrete strength and carbonization depth were done. The test results are recorded in Tables 13 and 14 respectively.

Fig. 5. Concrete pillar wrapped with the YHM-L/204 curing membrane

Table 13. Compressive strength of concrete pillars at different age cured by different curing method

Curing method	3d (MPa)	7d (MPa)	14d (MPa)	28d (MPa)	60d (MPa)	90d (MPa)
Natural curing	19.2	24.2	30.4	34.4	39.1	43.0
Conventional curing	19.9	26.3	35.4	39.5	44.4	48.7
Curing with YHM-L/204	20.5	27.4	37.1	42.0	47.8	51.6

Table 14. Carbonization depth of concrete pillars at different age cured by different curing method

Curing method	3d (mm)	7d (mm)	14d (mm)	28d (mm)	60d (mm)	90d (mm)
Natural curing	0	0	1	2.0	3.6	5.0
Conventional curing	0	0	0.6	1.3	2.6	3.6
Curing with YHM-L/204	0	0	0	0.5	1.4	2.0

From Table 13, it can be seen that during the whole test period, the strength growth of the concrete pillar cured with the YHM-L/204 curing membrane was higher than those under the other two curing methods. At 90d age, compared with the natural curing, strength of the concrete pillar under the conventional curing increased by 5.3 MPa, while strength of the concrete pillar under curing with the YHM-L/204 curing membrane increased by 8.6 MPa, and reached 51.6 MPa. Therefore, the use of YHM-L/204 curing membrane promoted the continued hydration in the concrete pillar and improved strength of the concrete pillars at later age.

As shown in Table 14, in the whole test period, carbonization rate of the concrete pillar under curing with the YHM-L/204 curing membrane was much lower than those under the other two curing methods. At 28d age, the carbonization depth of the concrete pillar under curing with the YHM-L/204 curing membrane was only 0.5 mm. At 90d age, compared with the natural curing, the carbonization depth of the concrete pillar under the conventional curing was reduced by 28%, while the carbonization depth of the concrete pillar under curing with the YHM-L/204 curing membrane was only 2.0 mm and reduced by 60%. Therefore, the use of YHM-SL/204 curing membrane bettered the concrete surface environment, and effectively improved the concrete resistance to carbonation.

5 Construction of Concrete Curing with New-Type Curing Membrane

5.1 Curing of Horizontal Surfaces of Concrete

Cement concrete is still plastic before the initial setting. At this time any pressure on concrete will produce irreversible deformation, so the new-poured concrete should not be covered for curing before the completion of the initial setting. Otherwise, it will make surface rough, having an effect on the apparent quality. On the other hand, the concrete should have joint-cutting after the final setting. At this time, if the concrete is covered with the curing membrane, the curing membrane has to be removed, or it will be cut into small pieces by joint-cutting which will worsen its effect of retaining moisture. In order to ensure the normal hydration of concrete before the final setting, the concrete was sprayed with water or curing agent for curing. When the joint-cutting of the concrete was done, the debris on the joints and the slabs was promptly cleaned up. And then water was sprayed on the pavement until the free water on the pavement surface reached 3–4 mm deep, which would leave adequate amount of water for the later-placed curing membranes with the polymer absorbent materials to absorb. After that, the curing membranes were spread over the pavement (as shown in Fig. 6). Note that the two adjacent curing membranes were overlapped with a width of no less than 5 cm. The edge of the curing membrane was equipped with the lap adhesive, which could make the preliminary adhesion between the overlapped edges of the two adjacent curing membranes. And then Scotch tape of 5 cm width was used to enhance this kind of adhesion. In order to prevent the curing membranes from being blown open by the wind or washed away by heavy rain during the whole curing period, sandbags of about 1 kg weight or cement concrete blocks or aggregates were placed on the 10-cm-wide edge of curing membrane. Note that the curing membranes covered the sidewalls of the concrete slabs completely. During the 28d curing, no more water was sprayed.

The major steps of construction for curing the cement concrete pavement with the new-type water-saving and moisture-retaining curing membrane were summarized as follows: (1) to spray enough water over the concrete pavement; (2) to place the new-type curing membranes; and (3) to process the joints between the curing membranes (gluing and placing heavy objects).

Fig. 6. Construction of curing cement concrete pavement with the new-type curing membranes

5.2 Curing of Vertical Surfaces of Concrete

The major steps of construction for curing the vertical surfaces of cement concrete with the new-type water-saving and moisture-retaining curing membrane (as shown in Fig. 7) were listed as follows: (1) to demold the concrete 24 h later after casting, and remove the floating slurry, sand and other debris from the concrete surface so as to ensure the concrete surface completely clean; (2) to place the curing membrane exclusive for curing the vertical surfaces onto the concrete surface and control the lap width to be no less than 50 mm between the two adjacent curing membranes; (3) to water continually from top to bottom at joints of the two adjacent curing membranes, and ensure watering to be slow, uniform and moderate till the polymer absorbent material within the curing membrane visibly swelled to 3–5 mm thick (After the initial watering, no more watering was needed during the whole curing period); and (4) to seal the joints of the two adjacent curing membranes with tape.

6 Conclusion

A new-type water-saving and moisture-retaining curing membrane was developed by the research of this paper after a series of tests for water absorption, moisture retention and degradation of the controllable polymer absorbent materials. Furthermore, when the new-type curing membrane was used for curing cement concrete, temperature and humidity of the concrete under curing were tested. And cracking, strength and carbonation depth of the concrete cured by different methods were also tested. Based on the results of the above tests, the following conclusions can be drawn:

(1) When the new-type water-saving and moisture-retaining curing membrane developed in the research of this paper was used for curing concrete, the amount of water needed for curing was only 1/(50–100) of that under curing with sacks or geotextile, and 1/10 of that under curing with agricultural film. In the place short of water, the new-type water-saving and moisture-retaining curing membrane is a better choice for concrete curing.

Fig. 7. Construction of curing concrete pillar with the new-type curing membranes

(2) At the same curing age and the same outside temperature, the use of the new-type water-saving and moisture-retaining curing membrane developed in the research of this paper significantly reduced the number of harmful and extremely harmful pores in concrete, which would improve the durability of concrete.
(3) The use of the new-type water-saving and moisture-retaining curing membrane developed in the research of this paper reduced the temperature difference between the inside and the outside of the concrete so as to lessen the occurrence of the microcracks in concrete at early age caused by the temperature difference.
(4) The new-type water-saving and moisture-retaining curing membrane developed in the research of this paper could improve strength and carbonation resistance of concrete.

Acknowledgements. This research was supported by the National Natural Science Foundation of China under Grant 51578080 and Grant 51178064.

References

Guo, B.L., Kong, X.M.: Effect of moisture-curing method on microstructure of concrete. China Concr. **16**(10), 54–57 (2010). (in Chinese)

Huang, J., Yang, X.H.: Curing effect of new-type water-saving and moisture-retaining curing membrane. Constr. Des. Eng. **5**, 48–49 (2004). (in Chinese)

JTG F80/1-2004: Quality Inspection and Evaluation Standards for Highway Engineering, Section 1 Civil Engineering. Beijing: China Communications Press (in Chinese) (2004)

Khatib, J.M., Mangat, P.S.: Influence of high-temperature and low humidity curing on chloride penetration in blended cement concrete. Cem. Concr. Res. **32**(11), 1743–1753 (2002)

Kong, X.M., Li, Q.H.: Influence of super absorbent polymer on dimension shrinkage and mechanical properties of cement mortar. J. Chin. Ceram. Soc. **37**(5), 855–861 (2009). (In Chinese)

Nevile, A.M.: Properties of Concrete, 4th edn. Wiley, New York (1996)

Ramezanianpour, A.A., Malhotra, V.M.: Effect of curing on the compressive strength, resistance to chloride-ion penetration and porosity of concretes incorporating slag, fly ash or silica fume. Cement Concr. Compos. **17**(2), 125–133 (1995)

Wu, Z.W., Lian, H.Z.: High Performance Concrete. China Railway Publishing House, Beijing (1999). (in Chinese)

Yang, W.D., et al.: Technical and economic performances of curing agent of high performances. Low Temp. Architect. Technol. **4**, 12–13 (2004). (in Chinese)

Yang, Z., Huang, J., Zhao, J., Yao, J.L.: Application of moisture-retaining curing membrane in cement concrete pavement construction. Hunan Transp. Sci. Technol. **34**(3), 28–29 (2008). (in Chinese)

Yokoyama, K., Hiraishi, S., Kasai, Y., Kishitani, K.: Shrinkage and cracking of high strength concrete and flowing concrete at early ages. In: Malhotra, V.M. (ed.) Proceedings of the 4th CANMET/ACI International Conference on Superplasticizers and other Chemical Admixtures in Concrete. ACI SP-148. American Concrete Institute, pp. 243–258(1994)

Characteristics of Excavation-Induced Deformation Associated with Different Propped Retaining Walls in Soft Soil

Nian-wu Liu[1], Feng Yu[1(✉)], and Xiao-nan Gong[2]

[1] School of Civil Engineering and Architecture, Zhejiang Sci-Tech University, Hangzhou, China
pokfulam@zstu.edu.cn
[2] Research Center of Coastal Urban Geotechnical Engineering, Zhejiang University, Hangzhou, China

Abstract. This paper reports the field performance of deep excavation of two subway-station cases, including the lateral wall deflection and the settlements of surrounding soil. The involved retaining structures were contiguous pile wall (CPW), soil-mixing wall (SMW) and diaphragm wall (DW), all embedded in soft clay. The profile of the measured wall deflections showed a typical bulging type at the end of the excavation. The ratios of the measured maximum wall deflection to the excavation depth were found to be similar among the three types of retaining walls. The relationship between the normalized maximum wall deflection and the factor of safety was described. The observed maximum deflections of the retaining walls were generally around the lower bound proposed by existing findings. This may be attributed to the use of prestressed struts and the short horizontal span of the excavation. The measured settlement of ground surface increased linearly with the increasing maximum lateral wall deflection; meanwhile, the magnitude of settlement became extraordinarily large due to the presence of sludgy soil. A concave pattern was proposed for the surface settlement profiles for the three types of retaining walls. The concave-pattern profiles were proposed for the subsurface settlement at the depths of 0, 5 and 10 m while the spandrel-pattern profiles were proposed at the depths of 15, 20 and 25 m. The critical depth derived from the concave-type to the spandrel-type was 63% of the excavation depth. As to the contiguous pile walls and the soil-mixing walls, the scopes of the involved area influenced by excavation were different.

Keywords: Field test · Excavation · Site investigation · Soft soil

Notations

H_e	Excavation depth
L	Distance from the excavation corner
x	Distance from the retaining wall
δ_{hm}	Maximum wall deflection
δ_{vm}	Maximum ground surface settlement
δ_v	Ground surface settlement

W. Steyn et al. (eds.), *Pavement Materials and Associated Geotechnical Aspects of Civil Infrastructures*, Sustainable Civil Infrastructures,
https://doi.org/10.1007/978-3-319-95759-3_2

δ_h Wall deflection
H_w Depth of wall
h_n The spacing of the struts

1 Introduction

Deep excavations are commonly used to utilize the urban space. Vertical retaining walls supported internally by multiple prop levels were used to open up the desired underground space and avoid the damage to adjacent structures or buildings. Several empirical or semi-empirical methods have been proposed to predict the induced ground settlement (Peck 1969; Clough and O'Rourke 1990; Long 2001; Moormann 2004; Liu et al. 2005; Wang et al. 2005; Tan and Wei 2012). For practical use, proper wall type should be chosen to decrease the wall deflections and ground movements. Therefore, there is a need to propose an accurate analysis of wall deflection and ground surface settlement of different supporting structures. Based on field and numerical results, many researchers found that the ground response to excavation was significant affected by the dimensions of excavation and the height of retaining wall (Bono et al. 1992; Wong and Patron 1993; Ou et al. 1993, 2000a, b; Lee et al. 1998; Finno and Bryson 2002; Finno and Calvello 2005; Finno et al. 2007).

However, the past researches have not carried on detailed analysis on the types influence of three-dimensional retaining wall. Contiguous pile wall (CPW), soil-mixing wall (SMW) and diaphragm wall (DW) supported internally by multiple prop levels were commonly used in deep excavation. To get a better understanding of the deformation characteristics of the three types retaining wall, detailed measured data of two subway station excavations using the three types wall were analyzed. The main purpose are, (1) to clarify the relationship between the settlement of ground surface and the maximum lateral deflections of three types wall; (2) to investigate the surface settlement profiles of three types; (3) to analyze the subsurface settlement profiles of the CPW and SMW at different depths.

2 Brief Description of the Case Histories

Two subway station (station 1 and station 2) excavations were approximately 20, 240 m in plan with excavation depth of 14–17 m. The plan view and the three types retaining wall are shown in Figs. 1 and 2. The contiguous pile wall and soil-mixing wall were used in station 1, while the diaphragm wall was used in station 2. Three different types retaining wall, i.e., CPW, SMW and DW with 4 levels strut were adopted in this case. The bottom-up method was employed in both of the excavations. When the construction of retaining walls was finished, the excavation and strutting systems were going to be executed. The magnitude of over-excavation depth in both of the stations was restricted within 0.5 m.

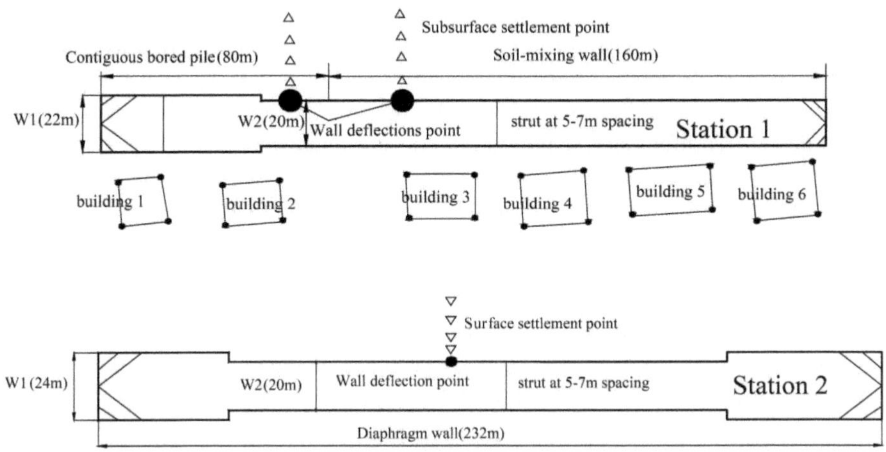

Fig. 1. Plan view showing a typical station and arrangement of strutting system

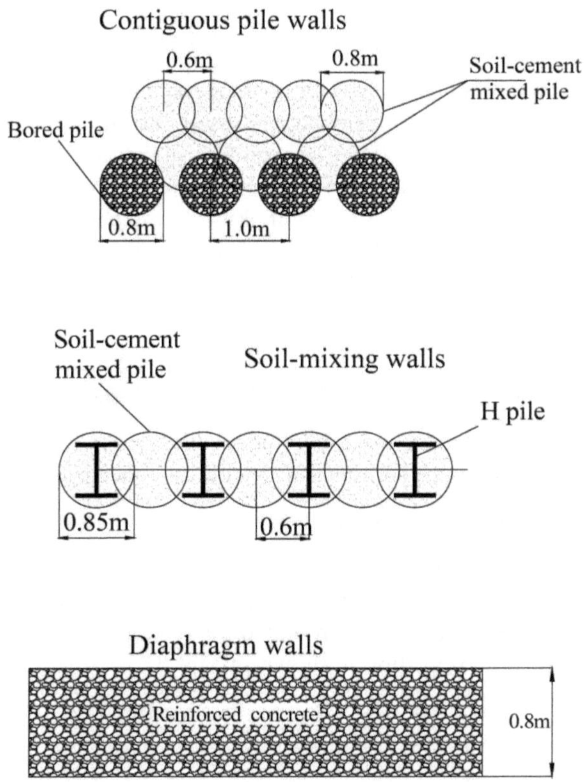

Fig. 2. Three types of retaining wall

A few kinds pipe strut with outer diameters of 0.4, 0.58, 0.609 and 0.8 m, respectively, are commonly used in China. Wall thicknesses are chosen to correspond to these diameters. When the type of pipe strut was chosen, the parameters would be input into the LIZHENG software. Then the outputs are checked out and compared with the design requirements. If the outputs do not meet the design requirements, the inputs will be changed until the outputs meet the design requirement. There was a positive correlation between the force and the deformation of the struts. So adjustable connectors and hydraulic jack were installed at the tips of the steel pipes for pre-stressing purpose. The magnitude of the pre-stress value was obtained from the experience of similar project. In this project, the upper two levels of struts were concrete beams with height of 1.2 m and width of 0.8 m, while the bottom two levels were steel pipes with thickness of 16 mm and outer diameter of 0.609 m.

The construction sequence of the CPW is summarized as below:

(1) Two rows of soil-cement mix piles with 0.8 m diameter and pile 0.6 m spacing were installed. The row spacing was 1.0 m.
(2) Concrete bored piles with 0.8 m diameter and 1.0 m pile spacing were installed. There was an 0.2 m overlap between the concrete piles and previously installed soil-mixing piles.

The construction sequence of the SMW is summarized as below:

(1) Soil-cement mixing piles with 0.85 m diameter and 0.6 m pile spacing were installed. Ordinary Portland cement was used, in addition, a weight ratio of cement to soil was chosen at 0.2. Concerning installation of the piles, cement was injected into soil which had been heavily disturbed by the stir, so as to form vertical columns of mixtures. A series of physical and chemical reactions occurred when the cement was mixed with the soil. The strength of the piles can be improved by increasing the numbers of stir and the weight ratio of cement to soil.
(2) A steel H-pile was immediately inserted into a newly installed soil-cement mixing pile prior to the installation of a next column.

The construction sequence of the DW is summarized as below:

(1) A trench was excavated by using bentonite slurry.
(2) Steel cages were put into the trench followed by concreting.

The excavation sequence is described as follow:

(1) Three types supporting structure were installed.
(2) The struts were installed by step as the excavation was being proceeded. Their corresponding depths below the ground were 1, 4.2, 7.6, and 12.2 m. A typical cross section of the excavation is shown in Fig. 3.

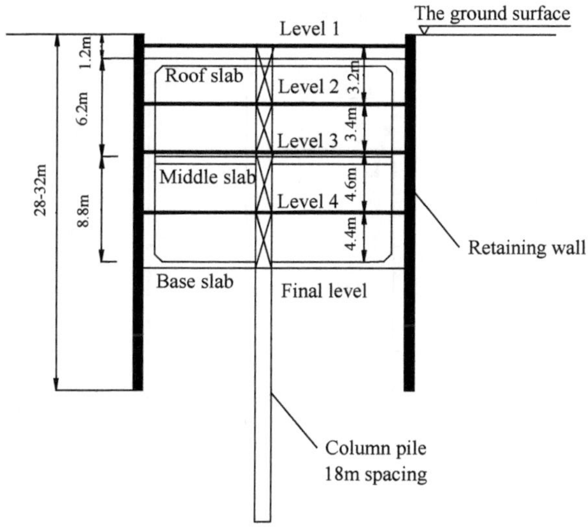

Fig. 3. Cross section of a typical excavation

3 Ground Profiles

The excavation site was located in Hangzhou, China, and had high groundwater table and thick soft soil layer. A series of geotechnical explorations including in-situ tests (e.g., standard penetration test and cone penetration test) and laboratory tests were conducted. 18 boreholes were done in both stations. The spacing of the adjacent boreholes was 30 m and the borehole was located in the position of the retaining walls. Figure 4 shows a typical soil profile and associated soil properties in the site. The effective cohesion, c', was determined by consolidated undrained triaxial tests, generally saying, the strength envelope of normally consolidated soil passed through the original point of coordinates, resulting in higher "apparent" friction angle. Actually, the density of soil increases under loading. Thus, the soil exhibits lower "true" friction angle and cohesion greater than zero. Whereas the compressibility modulus, $E_{0.1-0.2}$, was derived from odometer tests which were under pressure ranging from 100 to 200 kPa. Six layers were classified according to the soil properties. The top layer was a thin, artificial fill. Both the second and fourth layers were soft muddy clay including mica, moreover two layers contained other organic materials that are high void ratio and high water content. The third and fifth layers were brownish yellow, soft to medium plastic clay. Note that the second and the third layers were slightly over-consolidated. The sixth layer was stiff silty clay with thickness of 10–25 m. In station 1, the natural groundwater table was generally about 1 m below the ground surface. In station 2, the natural groundwater table was generally about 2.2 m below the ground surface. The groundwater level was lowered down to 4–6 m during excavation. The saturated permeability of the clay was between 10^{-6} and 10^{-4} m/s. Furthermore, a layer of sludgy soil with thickness ranging from 2 to 4 m was occasionally found at the location of station 2. The provided SPT results were raw blow counts and they were an average of 6 tests conducted in each station.

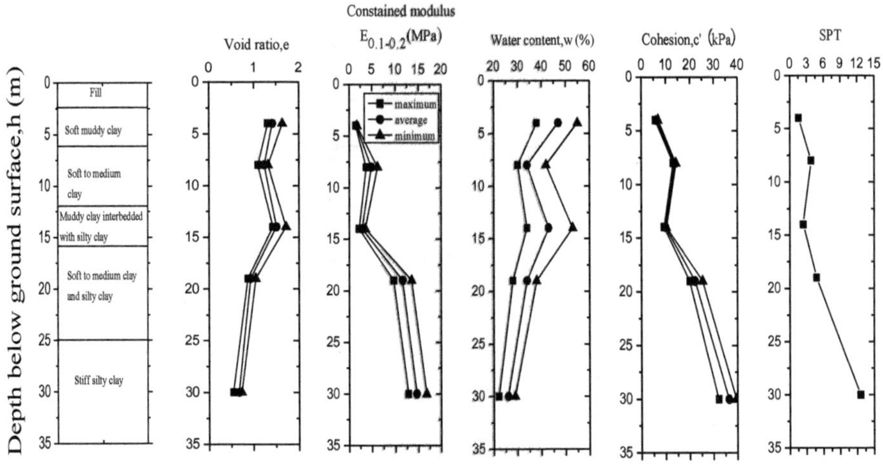

Fig. 4. Soil profile and geotechnical parameters at the test site

4 Instrumentations

To observe a performance of the three types of retaining walls embedded in thick soft clay, a comprehensive instrumentation diagram was conducted. The deflections of the wall were measured using 24 inclinometer tubes in station 1 (8 inclinometer tubes and 16 inclinometer tubes were installed in the CPW and the SMW, respectively.) and 22 inclinometers were installed in station 2. The inclinometers were installed in four sides at 15–20 m intervals. The wall deflection measuring point was 0.5 m intervals along its depth. Lateral displacements of the top of the wall were also measured. 4 or 5 settlement survey points perpendicular to the wall with the same locations to the inclinometer tubes were installed to investigate the settlement of the ground surface. The reference point was located at a distance of about five times the final excavation depth, H_e, from the excavation site. A magnetic extensometer system was also used to investigate the subsurface settlements at different depths in the contiguous pile walls section and soil-mixing walls section and the surface settlement was chosen as reference point.

5 Observed Performance of the Excavation

The wall deflection, surface settlement, subsurface settlement and building settlement were measured during the whole process of construction. For simplicity, only part of typical field data was presented and analyzed. Notations of the deformation-related parameters are shown in Fig. 5.

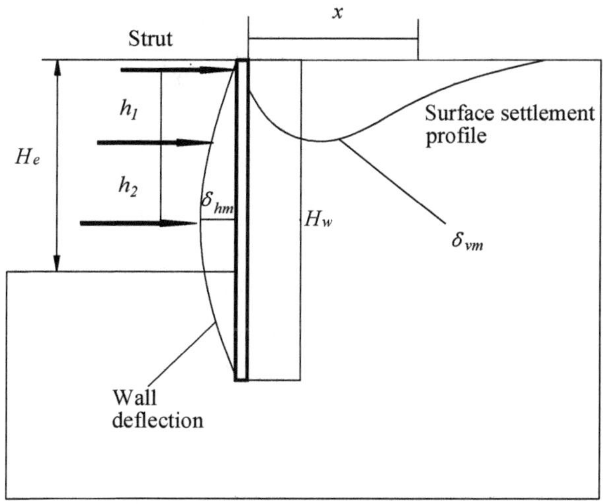

Fig. 5. Definition of excavation variables

5.1 Wall Deflections

Figure 6 shows typical wall deflections of the three types retaining wall. Note that the retaining walls were supported by 4 levels of strut among which the upper two levels were concrete struts and the lower two levels were steel struts. Inclinometer readings of the wall were calibrated by the measured lateral displacements at the top of the wall. Figure 6 illustrates that the deflection profiles associated with the three types retaining wall are similar and develop into a bulged profile as the excavation depth increased. The depth of maximum wall deflection rises with increasing excavation depth. The pile toe displacements are found to be close to zero at the excavation depth of 16 m. Wang et al. (2010) suggested that the toe movement in soft soil may exist if the wall toes were not embedded in competent stratum. However, the toe movement was small in the soft soil if the pile length was large.

Figure 7 plots the relationship between the normalized maximum wall deflection δ_{hm}/H_e and the factor of safety (FOS). In Fig. 7, FOS is calculated using the method of Bjerrum and Eide (1956), δ_{hm} is the maximum wall deflection, and H_e is the excavation depth. Moreover, the field results of δ_{hm}/H_e for the three types retaining wall are also given in Fig. 7.

Figure 7 shows that δ_{hm}/H_e of the CPW varies between 0.17 and 0.29%, simultaneously its mean value is 0.24%. The δ_{hm}/H_e of SMW is in a range from 0.19 to 0.32% and the average value is 0.26%. δ_{hm}/H_e of the DW varies between 0.22 and 0.38% and the mean value is 0.30%. The mean values of δ_{hm}/H_e are slightly larger than the finding of Leung and Ng (2007) who suggested a value of 0.23%. However, these values are smaller than that of diaphragm walls installed in similar ground condition in Taipei and Singapore (Ou et al. 1993, 1998; Long 2001). There is no obvious

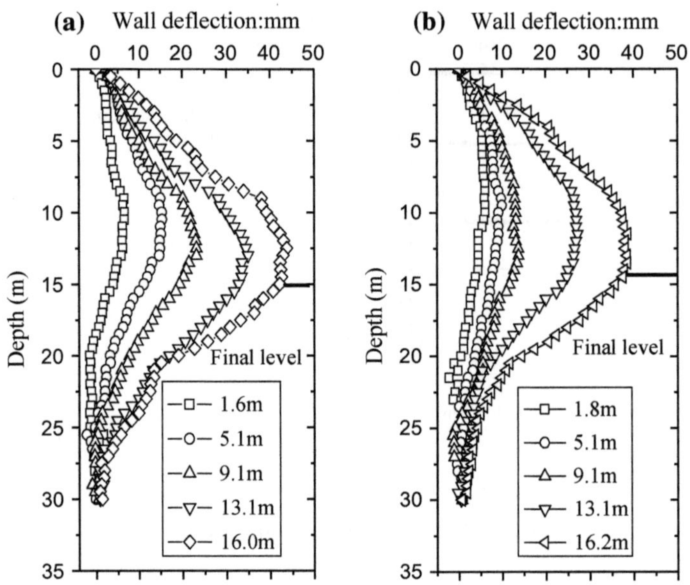

Fig. 6. Typical wall deflections at various excavation stages: **a** contiguous bored pile; **b** soil-mixing wall; **c** diaphragm wall

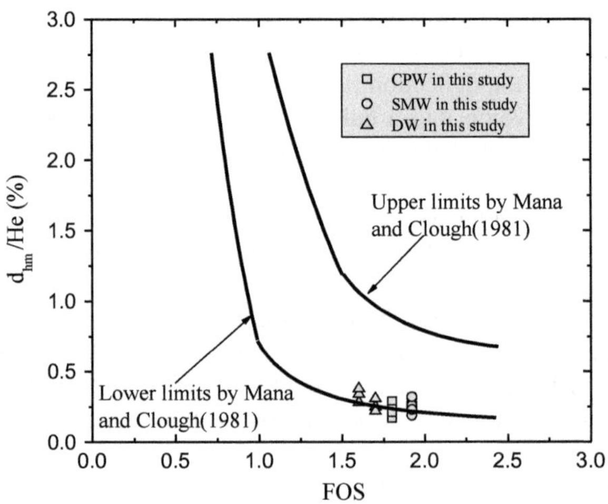

Fig. 7. Normalized maximum wall deflections of the three types of retaining walls versus FOS against basal heave

difference in the distributions of the δ_{hm}/H_e for the three types retaining wall. The observed δ_{hm}/H_e of the three different types of retaining walls is generally around the lower bound proposed by Mana and Clough (1981). This may be attributed to the use of prestressed struts and the short horizontal span of the excavation.

5.2 Relationship Between Maximum Lateral Wall Deflection and Ground Surface Settlement

Figure 8 presents the relationship between the measured maximum ground-surface settlement and the maximum lateral deflection of the three types of walls. It shows that the measured surface settlement increases with an increase in the maximum lateral wall deflection and appears to be linear with high reliability. This figure also shows that the measured data for the CPW and the SMW accord with two close lines defined by $\delta_{vm} = 1.35\delta_{hm}$ and $\delta_{vm} = 1.76\delta_{hm}$, respectively. This may be due to the difference in pile formation technology. Moreover, the relationship between the measured settlement of ground surface and the maximum wall deflection for the three types retaining wall is different from the findings of Mana and Clough (1981) and Wang et al. (2005). Mana and Clough (1981) suggested that the measured results of soil-mixing wall lay between $\delta_{vm} = 0.5\delta_{hm}$ and $\delta_{vm} = 1.0\delta_{hm}$. Wang et al. (2005) presented the measured results of soil-mixing wall lay between $\delta_{vm} = 0.2\delta_{hm}$ and $\delta_{vm} = 0.6\delta_{hm}$. Such difference may be attributed to the existence of the sludgy soil which gives rise to considerable surface settlement during excavation.

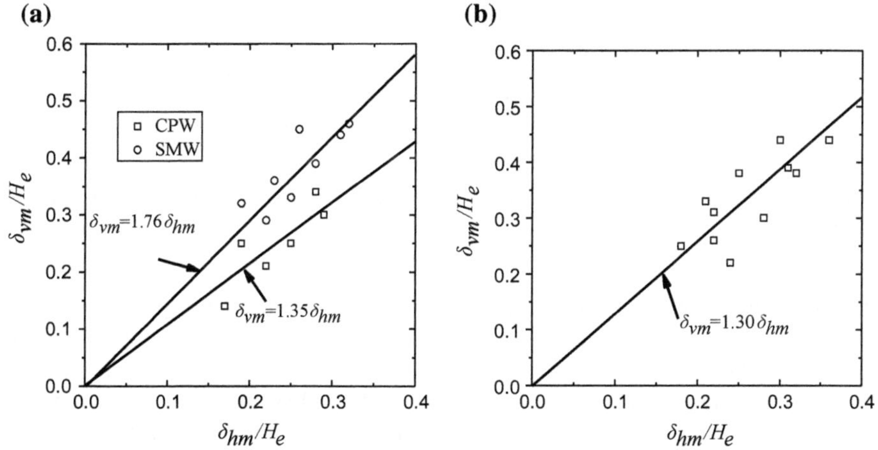

Fig. 8. Maximum ground settlement (δ_{vm}/H_e) versus maximum lateral wall deflection (δ_{hm}/H_e): **a** CPW and SMW at station 1; **b** DW at station 2

5.3 Ground Surface Settlement for Different Types of Walls

Surface-settlement profile around excavation is an important document used for assessing the potential damage to surrounding facilities The Surface-settlement profile was shown in Fig. 9 and compared with past findings. The measured maximum surface settlements for the CPW, the SMW and the DW are respectively 62, 59 and 88 mm. The observed settlements can be cataloged to Zone I which was suggested by Peck (1969). The relatively small settlements are related to several favorable factors such as: higher system stiffness, prestressing of the inner struts and soil improvement in the passive zone.

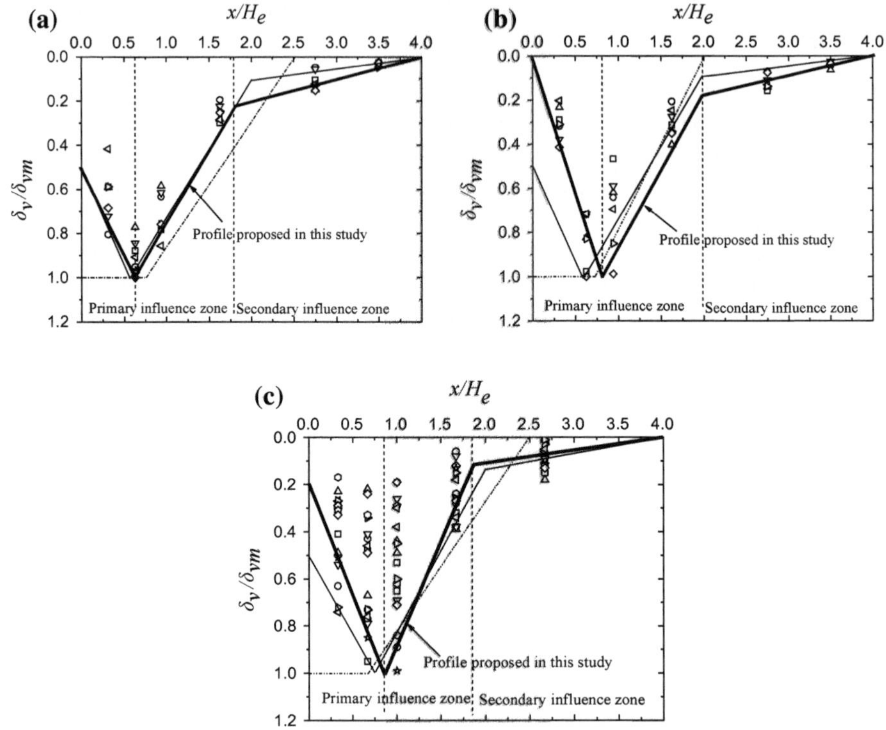

Fig. 9. Proposed surface-settlement profiles for different types of retaining walls: **a** CPW; **b** SMW; **c** DW

Figure 9 shows that the surface settlement profiles of different types retaining wall all follow a concave shape. The surface settlement profiles were presented as three types line. The heavy line represents as the profiles in this study, the fine line represents the findings of Hsieh and Ou's profiles (Hsieh and Ou 1998), the dashed line represents the findings of Clough and O'Rourkers' profiles (Clough and O'Rourke 1990). The surface settlement profiles can be divided into two zones (Hsieh and Ou 1998), i.e., the primary influence zone and the secondary influence zone. The two zones are divided by the critical value of x/H_e (see Fig. 3 for the definition of x), and the critical value is 1.6 for the CPW, 2.0 for the SMW and 1.7 for the DW. The primary influence zone can be further divided into two zones, i.e., zone I and zone II. In zone I, the value of δ_v/δ_{vm} increases with growing x/H_e and reached 1 at a certain distance. The value of δ_v/δ_{vm} decreases with an increase in x/H_e in zone II. Note that the decrease rate of δ_v/δ_{vm} in zone II of the primary influence zone is greater than that in the secondary influence zone. The maximum surface settlements for the two types wall used in station 1 are close. However, the values of x/H_e corresponding to the maximum surface settlements are different, i.e., 0.63 for the CPW and 0.81 for the SMW. The range of the primary influence zone for the CPW is also smaller than that for the SMW. Hashash and Whittle (1996), Whittle and Hashash (1994) suggested that the shape of the surface settlement profiles was largely affected by three factors: (1) the assumed roughness of the soil-wall interface that affects the soil movement; (2) the modeling of soil nonlinearity at small shear strain which influences both the magnitude and the distribution of settlement; and (3) the plane strain geometry. Referring to Fig. 2, the soil-wall interfaces of the three structures are different, the interface roughness of the CPW and SMW is larger than that of the DW. Moreover, the differences in modeling of soil nonlinearity at small shear strain are due to the wall formation technology. The CPW and the DW employ boring methods while the SMW employs cement mixing method. It also can be observed that the maximum surface settlement of diaphragm wall in station 2 is 30% larger than that in station 1. This is due to the existence of a considerable layer of sludgy soil in station 2.

5.4 The Subsurface Settlement

Figure 10 shows the settlement profiles at different depths and the fitting curves of the maximum settlement and zero settlement (i.e., the position where the settlement can be ignored) with depth.

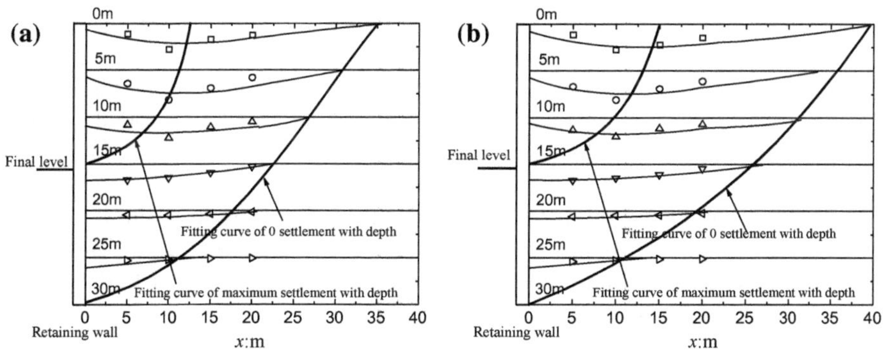

Fig. 10. Ground-settlement profiles at different depths: **a** CPW; **b** SMW

Figure 10 shows that both of the fitting curves of the maximum settlement and zero settlement with depth can be described using an upward parabola model with the starting point at the depths of 15 and 30 m. However, the scopes of the involved area influenced by the excavation are different. For the CPW, the intersection of the fitting curve of the maximum settlement and zero settlement with surface level are away from the retaining wall, the distances are 12.5 and 35 m, respectively. These values are smaller than that for the SMW with distances of 15 and 40 m, respectively. This is consistent with the statistical results of the surface settlement. This may be caused by the pile formation technology. It is well known the slurry-boring method in constructing contiguous pile walls will cause soil relaxing during the process of drilling. The settlement profiles behind the CPW and the SMW at different depth are similar. The concave-pattern profiles are proposed at the depths of 0, 5 and 10 m while the spandrel-pattern profiles are proposed at the depths of 15, 20 and 25 m. This is consistent with the findings proposed by Schuster et al. (2009). They suggested that the deflection of the wall toward the excavation zone were restricted at shallow depth and therefore proposed a concave-type profile. A spandrel-type was established because the wall deflected freely prior to the installation of struts. However, the critical depth ($0.63H_e$) derived from the concave-type to the spandrel-type is different from the findings of Schuster et al. (2009) who suggested that the critical depth was $0.25H_e$. Therefore, it is desirable to establish the lateral movement profiles at various depths for describing the response of building. As shown in Fig. 10, the maximum settlement occurred at the depth is 5 m, and the distance to the retaining wall is 10 m. This is consistent with the findings of Ou et al. (2000a). Although the maximum soil settlement occurred at the depth of 5 m, the width of the soil influenced by the excavation at 5 m is smaller than that of the surface soil.

6 Conclusion

Based on the detailed monitoring results of two multi-strutted excavations of subway stations in soft clay, the following conclusions can be drawn:

1. The deflection profiles of the three types of retaining wall are similar and develop into a bulged profile as the excavation depth increases. The toe movements are small despite the toe is embedded in soft clay. The distribution of the magnitude of δ_{hm}/H_e of the three types retaining wall are around the lower bound of curves proposed by Mana and Clough (1981) although part of the data exceed the lower limit.
2. The relationship between the measured settlement of ground surface and the maximum lateral wall deflection of the three types retaining wall is linear. The surface settlement in relation to the excavation is larger than other findings due to the existence of sludgy soil.
3. A concave pattern is proposed for the surface settlement profiles of the three types retaining wall. However, due to the differences in the soil-wall interface and the wall formation technology, the critical values of x/H_e between zone I and zone II and the scope of the primary influence zone are different from each other.
4. A concave-type profile is proposed for the subsurface settlement at the depths of 0, 5 and 10 m. A spandrel-type profile is proposed for the subsurface settlement at the depths of 15, 20 and 25 m. As to the contiguous pile walls and the soil-mixing walls, the scopes of the involved area influenced by excavation are different.

Acknowledgements. The authors would like to thank the funding supports from the research fund for commonwealth-orientated technology of Zhejiang Province (No. 2016C33020), the national natural science foundation of China (No. 51608485) and Zhejiang provincial natural science foundation (No. LZ17E080002), China. Great appreciation goes to the editorial board and the reviewers of this paper.

References

Bjerrum, L., Eide, O.: Stability of strutted excavations in clay. Géotechnique **6**(1), 32–47 (1956)
Bono, N.A., Liu, T.K., Soydemir, C.: Performance of an internally braced slurry-diaphragm wall for excavation support. In: Paul, D.B., Davidson, R.R., Cavalli, N.J. (eds.) Slurry Wall: Design, Construction and Quality Control, pp. 347–360. ASTM, Philadelphia (1992)
Clough, G.W., O'Rourke, T.D.: Construction induced movements of in situ walls. In: Lambe, P., Hansen, L.A. (eds.) Design and Performance of Earth Retaining Walls, pp. 439–470. ASCE, New York (1990)
Finno, R.J., Bryson, L.S.: Response of a building adjacent to stiff excavation support system in soft clay. J. Perform. Constructed Facil. **16**(1), 10–20 (2002)
Finno, R.J., Calvello, M.: Supported excavations: the observational method and inverse modeling. J. Geotech. Geoenviron. Eng. **131**(7), 826–836 (2005)
Finno, R.J., Bryson, L.S., Calvello, M.: Performance of a stiff support system in soft clay. J. Geotech. Geoenviron. Eng. **128**(8), 660–671 (2002)
Finno, R.J., Blackburn, J.T., Roboski, J.F.: Three-dimensional effects for supported excavations in clay. J. Geotech. Geoenviron. Eng. **133**(1), 30–36 (2007)
Hashash, Y.M.A., Whittle, A.J.: Ground movement prediction for deep excavations in soft clay. J. Geotech. Eng. **122**(6), 474–486 (1996)
Hsieh, P.G., Ou, C.Y.: Shape of ground surface settlement profiles caused by excavation. Can. Geotech. J. **35**(6), 1004–1017 (1998)
Lee, F.H., Yong, K.Y., Quan, K.C.N., Chee, K.T.: Effect of corners in strutted excavations: field monitoring and case histories. J. Geotech. Geoenviron. Eng. **124**(4), 339–349 (1998)
Leung, E.H.Y., Ng, C.W.W.: Wall and ground movements associated with deep excavations supported by cast in situ wall in mixed ground conditions. J. Geotech. Geoenviron. Eng. **133** (2), 129–143 (2007)
Liu, G.B., Ng, C.W.W., Wan, Z.W.: Observed performance of a deep multistrutted excavation in Shanghai soft clays. J. Geotech. Geoenviron. Eng. **131**(8), 1004–1013 (2005)
Long, M.: Database for retaining wall and ground movements due to deep excavations. J. Geotech. Geoenviron. Eng. **127**(3), 203–224 (2001)
Mana, A.I., Clough, G.W.: Prediction of movements for braced cut in clay. J. Geotech. Geoenviron. Eng. **107**(6), 759–777 (1981)
Moormann, C.: Analysis of wall and ground movements due to deep excavation in soft soils based on a new worldwide database. Soils Found. **44**(1), 87–98 (2004)
Ou, C.Y., Hsieh, P.G., Chiou, D.C.: Characteristics of ground surface settlement during excavation. Can. Geotech. J. **30**(5), 758–767 (1993)
Ou, C.Y., Liao, J.T., Lin, H.D.: Performance of diaphragm wall constructed using top-down method. J. Geotech. Geoenviron. Eng. **124**(9), 798–808 (1998)
Ou, C.Y., Liao, J.T., Cheng, W.L.: Building response and ground movements induced by a deep excavation. Géotechnique **50**(3), 209–220 (2000a)
Ou, C.Y., Shiau, B.Y., Wang, I.W.: Three-dimensional deformation behavior of the Taipei National Enterprise Center (TNEC) excavation case history. Can. Geotech. J. **37**(2), 438–448 (2000b)
Peck, R.B.: Deep excavation & tunneling in soft ground. State-of-the-art-report. In: Proceeding of 7th International Conference of Soil Mechanics and Foundation Engineering, ISSMGE, Mexico City, pp. 225–281 (1969)
Schuster, M., Kung, G.T.C., Juang, C.H., Hashash, Y.M.A.: Simplified model for evaluating damage potential of buildings adjacent to a braced excavation. J. Geotech. Geoenviron. Eng. **135**(12), 1823–1835 (2009)

Tan, Y., Wei, B.: Performance of an overexcavated metro station and facilities nearby. J. Perform. Constructed Facil. **26**(3), 241–254 (2012)
Wang, Z.W., Ng, C.W.W., Liu, G.B.: Characteristics of wall deflections and ground surface settlements in Shanghai. Can. Geotech. J. **42**(5), 1243–1254 (2005)
Wang, J.H., Xu, Z.H., Wang, W.D.: Wall and ground movements due to deep excavations in Shanghai soft soils. J. Geotech. Geoenviron. Eng. **136**(7), 985–994 (2010)
Whittle, A.J., DeGroot, D.J., Ladd, C.C., Seah, T.H.: Model prediction of anisotropic behavior of Boston blue clay. J. Geotech. Eng. **120**(1), 199–224 (1994)
Wong, L.W., Patron, B.C.: Settlements induced by deep excavations in Taipei. In: Proceedings of 11th Southeast Asian Geotechnical Conference, Singapore, pp. 787–791. The Institution of Engineers, Malaysia, Kuala Lumpur (1993)

Slope Stability Analysis of the Embankment Filled with River Sand and Weathered Mudstone Rock

Weidong Cao[1(✉)], Shutang Liu[1], and Yingyong Li[2]

[1] School of Civil Engineering, Shandong University, No. 17923 Jingshi Road, Jinan 250061, Shandong Province, People's Republic of China
{cwd2001, gtreesl}@sdu.edu.cn

[2] Shandong Province Transport Department, Highway Bureau, No. 19 Shungeng Road, Jinan 250002, Shandong Province, People's Republic of China
leeyyong@163.com

Abstract. To solve the shortage of land resources and the lacking of soil in expressway construction, the application feasibility of river sand and weathered mudstone rock (WMR) used for filling embankment with steep slope (slope ratio is 1:1) were evaluated in this paper. 24 kinds of embankment schemes in terms of the combinations of the river sand and WMR with different embankment heights were designed. The slope stability analyses of these embankments were conducted using numerical calculation method. The results show that the safety factors of river sand embankments with 3–4 m slope protection width (SPW) and composite embankment with 3 m SPW can meet the requirements, provided the embankment height is less than 8 m. While considering comprehensively the vertical deformation, location of potential sliding surface, and economic cost, the river sand embankment with 3 m SPW is the optimal scheme. When the embankment height exceeds 8 m, the increase of SPW has no obvious positive effect on the increase of safety factor.

1 Introduction

With the large-scale construction and development of expressway in China, the demand of filling material for embankment is growing, which often brings threat to the surrounding farmland beside the expressway. In the case of lacking farmland resources and strict protection of farmland in China, looking for other soils such as river sand or weathered mudstone rock (WMR) may be the unavoidable choice. The project case named Qing-Lin expressway goes through Linyi region of Shandong Province, China, where the farmland resources beside the expressway are scarcity and valuable. Since the cultivable land resource is not allowed to fill the expressway embankment, the alternative road materials must be considered. Fortunately, there exist rich river sand available near this expressway and a lot of minable WMRs in the surrounding mountains. So the section embankment of Qing-Lin expressway can be filled with river sand and WMR.

There have been some research and application of sand used in road embankments in China since 1990s [1–3]. In summary, the nature sand can be used to build road

W. Steyn et al. (eds.), *Pavement Materials and Associated Geotechnical Aspects of Civil Infrastructures*, Sustainable Civil Infrastructures,
https://doi.org/10.1007/978-3-319-95759-3_3

embankment covered edge with clay in China. In India, the natural sand is usually reinforced vertically by the geogrid and bamboo when used to fill subgrade. The test results indicate that this type of vertical reinforcement increases the bearing capacity and stiffness of sand subgrades [4]. Dune sand is stabilized with organic polymers for being used in road base in Saudi Arabia, and it can be improved quite significantly in compressive strength [5]. In Turkey sand is stabilized with fiber to reduce the brittleness and avoid a greater loss of shear strength of the sand body [6]. In Iran sand is mixed with shredded waste tires to improve the bearing capacity ratio (BCR) of sand. It is shown that the increasing of shred content increases the BCR. However, an optimum value for shred content is observed after that increasing shreds led to decrease in BCR [7]. Weathered rock is usually an appropriate road material, such as weathered granite has been often employed as roadbed material in China for many years [8, 9]. The decomposed granitic soil is widely used as the subgrade material in South Korea [10]. However few studies have been conducted for slope stability analysis of the expressway embankment filled with river sand and weathered mudstone rock.

Combined with the project case, the main objectives of this research are to conduct slope stability analysis for 24 kinds of expressway embankment schemes in terms of the combinations of the river sand and the WMR, and choose the most reasonable embankment scheme taking into account factor of safety, deformation and the economic factors as well.

2 Materials and Determination of Parameters

2.1 River Sand

The river sand was from construction site. The sieving curve of particle size distribution of the used river sand was shown in Fig. 1. The coefficient of uniformity (Cu) is 4.66, and the coefficient of curvature (Cc) is 0.95. According to China specifications [11], the river sand is classified as poorly graded soil. The maximum dry density and the optimum moisture content were obtained using modified compaction method, which were 1.87 g/cm^3 and 10%, respectively.

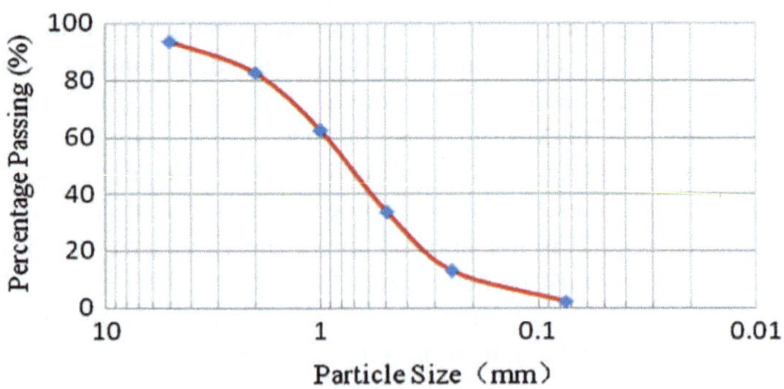

Fig. 1. Sieving curve of river sand

In order to conduct the numerical calculation of the embankment slope stability, the internal friction angle, cohesion, bulk modulus, shear modulus, poisson's ratio and other parameters also need to be determined. Shear strength parameters of river sand were determined by direct shear tests. The test results showed that the angle of internal friction of the river sand is 37°, and cohesion is 3.9 kPa when the compaction degree is 96% under the optimum moisture content. Based on indoor test and engineering experience, the elastic modulus of river sand was determined to be 40 MPa. According to relevant literatures [12, 13], poisson ratio and dilatancy angle of the river sand were taken as 0.4° and 15°, respectively. The bulk modulus and shear modulus can be calculated by formulas (1) and (2), respectively.

$$K = \frac{E}{3(1 - 2\mu)} \tag{1}$$

$$G = \frac{E}{2(1 + \mu)} \tag{2}$$

where, E is elastic modulus; μ is poisson's ratio; K is bulk modulus and G is shear modulus.

The calculated values of bulk modulus and shear modulus were 53 MPa and 15 MPa, respectively.

2.2 Weathered Mudstone Rock

The weathered mudstone rock (WMR) was also from the construction site. The sieving curve of particle size distribution of the used WMR is shown in Fig. 2. The Cu and Cc of the samples are 6.38 and 1.42, respectively. According to the two parameters, it can be judged that the WMR has good gradation. The maximum dry density and the optimum moisture content were determined by modified compaction method, which were 2.05 g/cm^3 and 9.8%, respectively.

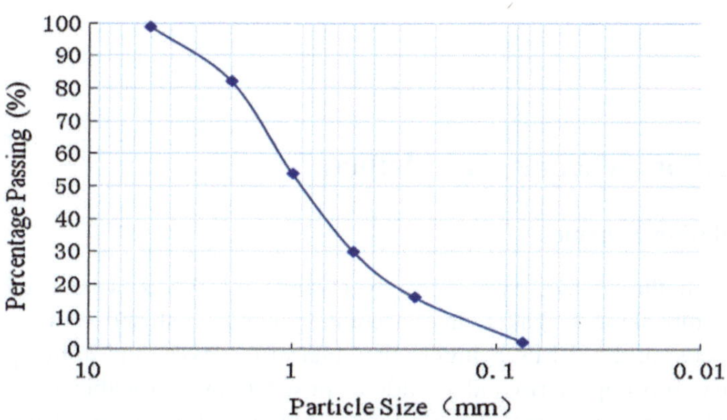

Fig. 2. Sieving curve of WMR

The shear strength parameters of the WMR were obtained using triaxial consolidated-undrained compression tests [11]. Based on the test results and project experiences, the internal friction angle and the cohesion were taken as 20° and 16 kPa, respectively. The volume modulus and shear modulus of the WMR were also calculated using Eqs. (1) and (2). According to the design drawing of the project case, the elastic modulus adopted was 60 MPa. Consequently, the calculated volume modulus and shear modulus were 100 MPa and 21 MPa, respectively. Finally, the dilatancy angle of the WMR was determined to be 8°.

2.3 The Foundation Soil Under the Embankment

According to the results from the geological survey, the foundation soil in the drilling range was, from up to down, silty clay with 5 m thickness, fully weathered argillaceous siltstone also with 5 m thickness, and strong weathered argillaceous siltstone with 10 m thickness. The determined parameters of strength and modulus by the department of the geological survey are listed in Table 1.

Table 1. Parameters of different soil layers of the foundation

Parameters	Soil layer 1 (silty clay)	Soil layer 2 (fully weathered argillaceous siltstone)	Soil layer 3 (strong weathered argillaceous siltstone)
Density (kg/m^3)	2042	1954	2006
Bulk modulus (MPa)	23	39	36
Shear modulus (MPa)	14	15	14
Internal friction angle (°)	10	25	35
Cohesion (kPa)	25	10	8
Dilatancy angle (°)	0	11	9

3 Calculation Schemes and Method

3.1 Calculation Schemes

Combined with the project case, both river sand and WMR were used for filling expressway embankment. Taken into account material cost and transport cost, the construction cost of the WMR is higher than that of the river sand in this project case. In order to try to occupy farmland as little as possible, as a probable solution, the 1:1 slope ratio of the embankment was designed, which is different from the commonly used value 1:1.5 recommended by specification. The design height of embankment involved in this research is from 8 to 12 m, and the design width on the top of the embankment is 34.5 m.

Comprehensively considering the slope stability, slope greening, and the construction cost, two groups of embankment schemes were designed in terms of the combinations of the river sand and the WMR with different embankment heights and slope protection widths (SPW). One group was pure river sand embankments with various SPW shown in Fig. 3, the other was composite embankments (filled with the river sand layer and WMR layer by turns) with various SPWs shown in Fig. 4. Among them, the thickness of each filling layer of river sand layer and WMR layer was 50 cm; the width of slope protection filled with WMR was taken as 0, 2, 3, and 4 m, respectively; the height of embankment was set as 8, 10, and 12 m, respectively. Therefore, a total of 24 kinds of calculation Schemes were designed.

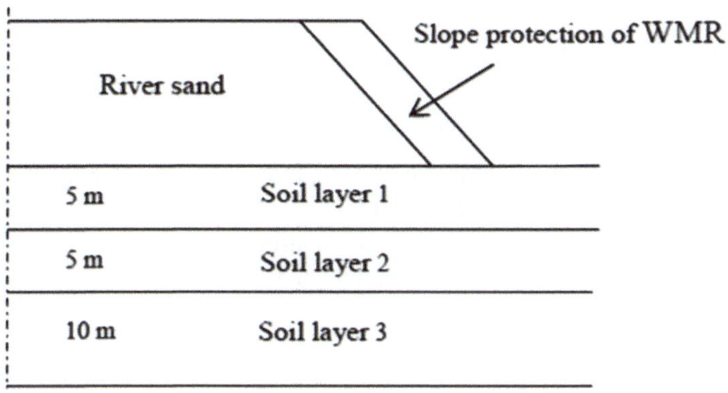

Fig. 3. Sketch of river sand embankment with slope protection

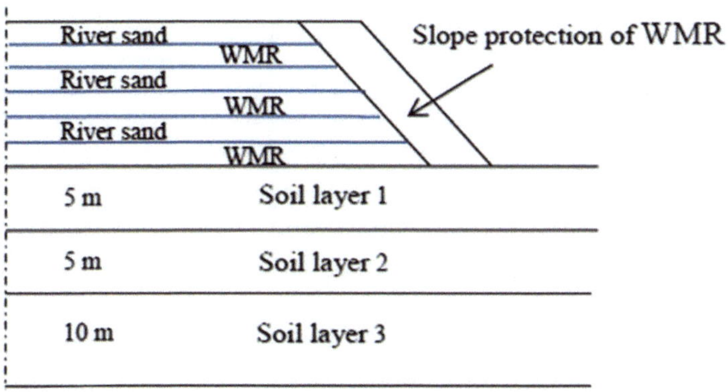

Fig. 4. Sketch of composite embankment with slope protection

3.2 Calculation Method

In this paper the strength reduction method (SRM) was employed to analyze the safety of slope and the deformation of embankment. This method is for an ideal elastic-plastic finite element system, in which the shear strength parameters of the sliding mass are gradually decreased by a varying reduction coefficient based on Mohr–Coulomb failure criterion, until the system achieves an unstable state and the finite element static analysis is not convergence or the slope fails, and the strength reduction factor at the situation is the safety factor [14, 15].

In the process of calculation, the strength parameters of the sliding mass and the sliding surface will be gradually reduced, that is, $c' = \frac{c}{F}$, and $\phi' = \arctan\frac{\tan\phi}{F}$.

Where, c is the cohesion and c' is the reduced cohesion; ϕ is the angle of internal friction and ϕ' is the reduced angle of internal friction; F is reduction coefficient or safety factor.

4 Results and Analysis

4.1 Calculation Results of Safety Factor and Analysis

Since river sand is cohesionless material, the slope of river sand embankment should be protected. In this research the WMR is chosen as the material of slope protection, whose appropriate width is determined by calculation and analysis. Figure 5 shows the trends of safety factor of river sand embankment with the increasing of SPW.

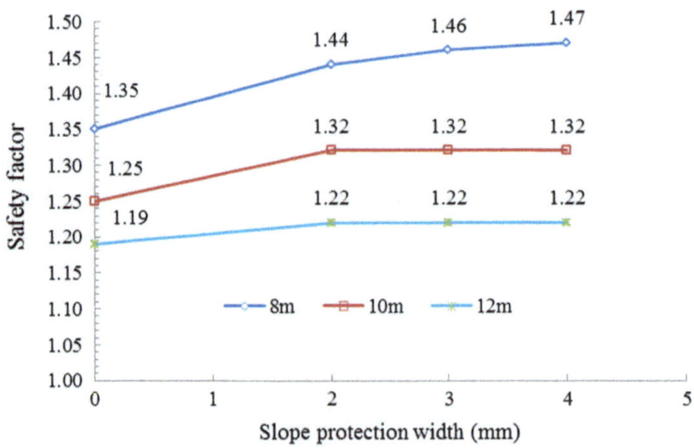

Fig. 5. Trends of safety factor of river sand embankment

Composite embankment is filled with both river sand layer and WMR layer, so it also needs protection. The varied SPW is the same as that of river sand embankment. The calculated results of values of safety factor for composite embankment are shown in Fig. 6.

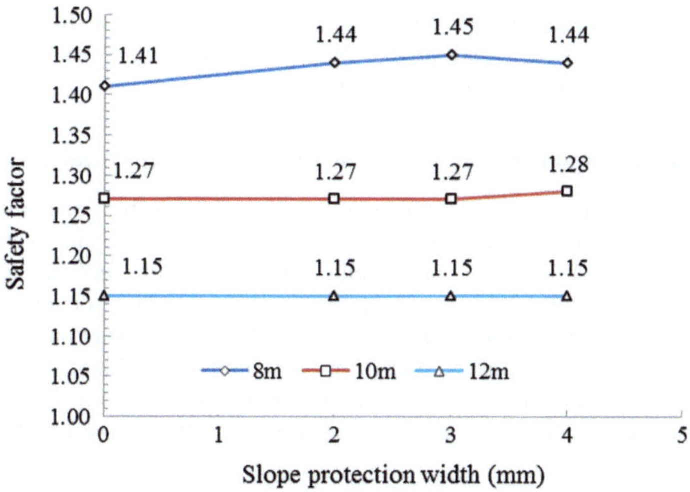

Fig. 6. Trends of safety factor of composite embankment

It can be from Figs. 5 and 6: (1) No matter what kind of material the embankment is made of, the values of slope safety factor of embankments decrease with the increase in embankment height, which is consistent with the common sense. (2) The effect of SPW on safety factor is related to the height of embankment. For river sand embankment, when the height of embankment is 8 m, the safety factor increases with the increase of SPW, but the increasing rate is gradually slowing down. When the height of embankment is 10 m, the safety factor first increases with the increase of SPW and then keeps a constant value. While when the embankment is 12 m high, the safety factor shows a similar trend as that of the embankment of 10 m height, but the increasing extent at the initial phase decreases. So the improvement of slope stability through the increasing of SPW is effective only when the height of embankment is not more than 8 m. When the embankment height reaches 10 m and higher, the increase in safety factor with the increase of SPW has been "eaten" by the decrease of safety due to the growing of embankment height. Therefore, when the embankment is higher than 8 m, the SPW can be determined from the point of view of functional protection and convenient for construction, such as 2.5 m. (3) For composite embankment, when the embankment height is 8 m, the safety factor of embankment slope increases gradually with the increase of SPW, reaching peak of 1.45 at the SPW of 3 m, and then decreases, which is obvious different from the trend of the river sand embankment. So, when the height of composite embankment is less than 8 m, the optimum SPW is 3 m. While when the embankment is higher than 10 m, the increase in SPW has no effect on slope safety factor. (4) Comparing the river sand embankment with composite embankment for the same height, the former shows more obvious effect of SPW increasing on safety factor.

According to engineering requirements [16], the minimum slope safety factor is 1.45. It can be concluded from Figs. 5 and 6 that river sand embankment at 8 m height with SPW of 3 m or 4 m, and composite embankment at 8 m height with 3 m SPW meet the safety requirement. The above embankments can be the feasible schemes. While when the embankments are higher than 8 m, decreasing the slope gradate may be an effective way to improve the slope safety.

4.2 Calculation Results of Embankment Deformation and Analysis

To select the optimum scheme of embankment from the feasible schemes, the maximum vertical deformation and the location of potential fracture plane or the width of the rupture body should be considered. The maximum vertical deformations (Z-displacement) of embankment include the maximum vertical depression and the maximum bulging deformations. It is obvious that the embankment with small deformation and narrow rupture body is priority recommended. The calculation results of embankment deformation for three feasible schemes are shown in Figs. 7, 8 and 9. The maximum deformation data and the values of safety factor are listed in Table 2.

(a) Nephogram of vertical deformation (b) Nephogram of strain shear increment

Fig. 7. Results of deformation for river sand embankment with 3 m SPW

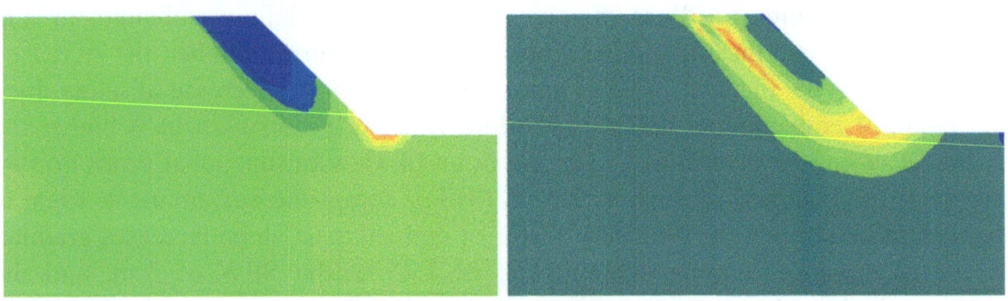

(a) Nephogram of vertical deformation (b) Nephogram of strain shear increment

Fig. 8. Results of deformation for river sand embankment with 4 m SPW

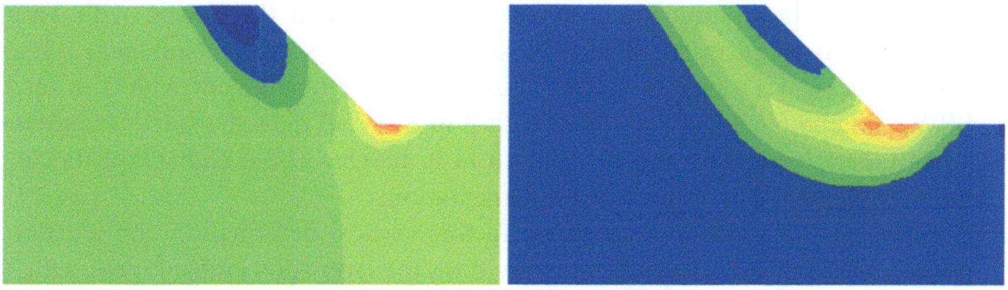

(a) Nephogram of vertical deformation (b) Nephogram of strain shear increment

Fig. 9. Results of deformation for composite embankment with 3 m SPW

Table 2. Maximum deformation data and safety factors of embankments

Kinds of embankment	SPW (m)	Maximum vertical depression (cm)	Maximum bulging deformation (cm)	Safety factor
River sand embankment	3	43.3	37.0	1.46
	4	39.8	40.2	1.47
Composite embankment	3	43.4	54.8	1.45

From the nephograms of vertical deformation of Figs. 7, 8 and 9, it can be seen that the locations of potential fracture plane or the size of rupture body of three schemes are different. The fracture plane of river sand embankment with 3 m SPW is outermost, the other river sand embankment with 4 m SPW moves further towards the inside of the slopes, while the location of potential fracture plane of composite embankment is in the innermost, and the depth of rupture body that embeds into the foundation increases in turn.

From Table 2, the depression deformation of river sand embankment with 4 m SPW is the smallest (39.8 cm), and the depression deformation of the other two embankments is both larger by 10% than 39.8 cm. River sand embankment with 3 m SPW has the smallest vertical bulging deformation (37.0 cm), which is smaller by 8% than that of river sand embankment, and smaller by 32% than that of composite embankment.

Therefore, based on the comprehensive consideration of vertical deformation, location of potential sliding surface, safety factor and economic cost, the river sand embankment with 3 m SPW is recommended as the optimal scheme.

5 Conclusions

On the basis of the calculation results and analyses of the embankment filled with river sand and weathered mudstone rock, the main conclusions are summarized as follows:

(1) In terms of the combinations of the river sand and the WMR with various embankment heights and SPWs, 24 kinds of expressway embankment schemes (slope ratio is 1:1) were deigned. The river sand embankment with 3 m or 4 m SPW, and composite embankment with 3 m SPW all meet the safety requirement when the embankments are less than 8 m in height. These embankments can be the feasible schemes in project implementation.
(2) Comprehensively considering safety factor, vertical deformation, the location of potential sliding surface, and economic cost, the river sand embankment with 3 m SPW is recommended as the optimal scheme among the feasible schemes.
(3) The effect of increasing SPW on slope stability has a close relationship with the height of embankment. The improvement of slope stability through the increasing of SPW is effective only when the height of embankment is not more than 8 m. When the embankments are higher than 8 m, decreasing the slope gradate may be an effective way to improve the slope safety.

It should be noted that the calculation results and some conclusions need further experimental verification.

Acknowledgements. This work was supported by Primary Research & Development Plan of Shandong Province in China (2016GSF116001) and the Transportation Department of Shandong Province, China, and the authors would like to acknowledge their financial support.

References

1. Li, S.Q., Xu, H.: Application research of Jiang Sha in construction of embankment. Highway **1999**(4), 21–24 (1999)
2. Jiang, X., Ling, J.M., Li, J.: Some critical problem on sand embankment design for expressway. Chin. J. Undergr. Space Eng. **2011**(7), 570–575 (2011)
3. Chen, Z., Li W.P.: Vibratory parameters and vibratory compacting mechanism of eolian sand. J. Changan Univ. (Nat. Sci. Edn.) **27**(1), 1–6 (2007)
4. Mandal J.N., Manjunath, V.R.: Bearing capacity of strip footing resting on reinforced sand subgrades. Constr. Build. Mater. **9**(1), 35–38 (1995)
5. Lahalih, S.M., Ahmed, N.: Effect of new soil stabilizers on the compressive strength of dune sand. Constr. Build. Mater. **1998**(12), 321–328 (1998)
6. Yetimoglu, T., Salbas, O.: A study on shear strength of sands reinforced with randomly distributed discrete fibers. Geotext. Geomembr. **21**(2), 103–110 (2003)
7. Hataf, N., Rahimi M.M.: Experimental investigation of bearing capacity of sand reinforced with randomly distributed tire shreds. Constr. Build. Mater. **20**(10), 910–916 (2006)
8. Li, X.Q., Jin Q.P.: Compacting deformation engineering characteristics of weathered soft rock mixture in subgrade. J. China Univ. Geosci. **19**(3), 298–306 (2008)
9. Guo, Y.C., Shen, A.Q., Gao, T., Li, W.G.: Assessment of weathering degree and road performance test of weathered rock as subgrade filling. J. Traffic Transp. Eng. **14**(3), 15–23 (2014)
10. Kim, D., Sagong, M., Lee, Y.: Effects of fine aggregate content on the mechanical properties of the compacted decomposed granitic soils. Constr. Build. Mater. **19**(3), 189–196 (2005)
11. China Standard JTG E40-2007: Test Methods of Soils for Highway Engineering. China Communications Press, Beijing (2007)

12. Zheng, H., Li, C.G., Lee, C.F., et al.: Finite element method for solving the factor of safety. Chi. J. Geotech. Eng. **24**(5), 626–628 (2002)
13. Chen, Y.M., Xu, D.P., FLAC/FLAC3D: Basic and Engineering Examples. China Communications Press, Beijing (2002); China Water and Power Press (2009)
14. Huang, M.S., Jia, C.Q.: Strength reduction FEM in stability analysis of soil slopes subjected to transient unsaturated seepage. Comput. Geotech. **2009**(36), 93–101 (2009)
15. Shen, H., Klapperich, H., Abbas, S.M., Ibrahim, A.: Slope stability analysis based on the integration of GIS and numerical simulation. Autom. Constr. **26**(10), 46–53 (2012)
16. China Specifications JTG D30-2015: Specifications for Design of Highway Subgrades. China Communications Press, Beijing (2015)

Investigating Effects of Individual Fracture Length on Behaviour of Weak Rock Using Discrete Element Method

Xiangyu Zhang(✉), Behzad Fatahi, and Hadi Khabbaz

School of Civil and Environmental Engineering, University of Technology, Sydney, Australia
Xiangyu.Zhang@uts.edu.au

Abstract. In this paper weak rock specimens with different individual fracture lengths are numerically simulated using the discrete element method (DEM). Effects of micro or macro-mechanical responses of intact and fractured specimens subjected to triaxial test have been studied. Various individual fracture lengths with a given fracture density within the weak rock specimens were reproduced using the particle flow code in three-dimension software (PFC3D). Different lengths of fractures were simulated by altering the size of each fracture to give insight over the influence of continual fractures and non-persistent fractures within bonded assemblies. As expected, for a given fracture density the individual fracture length affected the strength and deformability of rock mass. For an individual fracture length to specimen width ratio (the normalized fracture length) less than a limiting value, the effects of the individual fracture length on the stress-strain behaviour of rock specimens were more evident. Indeed, the strength decreased with decreasing the normalized fracture length. However, with a ratio above the limiting value, the effects of the individual fracture length were minimal. It can be concluded that for a given fracture density, present of shorter mini-fractures could be potentially more detrimental to stiffness and strength of the rock mass in comparison to longer major fractures.

1 Introduction

Weak rocks have complex characteristics including low shear strength, often being fractured and weathered with significant deterioration. These aspects contribute to the complexity of understanding the behaviour of weak rocks, especially for the tunnel and deep foundation designs in which difficult ground conditions are often encountered (Goodman and Ahlgren 2000; Klein 2001; Nickmann et al. 2006; Santi 2006). Many studies have been conducted on classifying and investigating the behaviour of weak rocks through the field, laboratory and numerical approaches (Goodman and Ahlgren 2000; Nickmann et al. 2006; Santi 2006; Sun et al. 2006). The effect of characteristics of joints in terms of the pattern, orientation, spacing, size, distribution, roughness, and aperture on various rock types have also been studied by many researchers (Bahaaddini et al. 2013; Chiu et al. 2013; Prudencio and Jan 2007; Tsoutrelis and Exadaktylos 1993). According to Klein (2001) and Santi (2006), discontinuities tend to contribute

W. Steyn et al. (eds.), *Pavement Materials and Associated Geotechnical Aspects of Civil Infrastructures*, Sustainable Civil Infrastructures,
https://doi.org/10.1007/978-3-319-95759-3_4

significantly to the weakness and deformability of rock masses. However, for a given fracture density, the effects of the individual fracture length on the weak rock mass is not understood well.

The discrete element method (DEM) based on particle mechanics has several advantages to investigate the effects of individual fracture length on the strength and deformability of weak rock. DEM allows for the explicit consideration of the heterogeneity and localized microscopic-mechanical properties of the rock mass and replicating the natural fractures (Bahaaddini et al. 2013; Chiu et al. 2013; Potyondy and Cundall 2004). The bonded particle model, which is embedded in the discrete element method, can replicate the processes of breaking the contact bonds for rocks, and the cementation degradation of cemented soils due to volumetric or shear strains (Nguyen and Fatahi 2016; Nguyen et al. 2017; Potyondy and Cundall 2004). In this paper, the rock mass has been simulated by adopting discrete element method to investigate the alteration of stress-strain behaviour due to the variation in the individual fracture length for a given fracture density in the rock sample.

2 Numerical Modeling

The numerical model was initially calibrated using the triaxial test results on a weak rock sample reported by Sun et al. (2006) for a confining pressure of 392 kPa. The weak rock in the study was described as weathered porphyrite obtained from a dam site in Hyogo Prefecture in Japan. The void ratio and specific gravity of this specimen were reported as 1.25 and 2.74, respectively.

2.1 Numerical Model Components

As recommended by Potyondy and Cundall (2004), to pull particles together to create an initial sample, first, a loose assembly of particles was generated randomly with the specified diameter adopting frictionless walls. The system was adjusted by allowing the particles to move considering no friction. Then, a nominal low isotropic stress was applied on the assembly of particles to reduce the magnitude of locked-in stresses. Finally, floating particles which had less than three contacts were eliminated, and then bonded particle model (BPM) was assigned as the contact model between particles. The rock specimen had a size of 100 × 100 × 200 mm. The median diameter of the particles adopted in the numerical model was 3.5 mm, while the ratio of maximum diameter to minimum diameter (d_{max}/d_{min}) was 2 with a uniform distribution, similar to those reported in the literature for investigating the behaviour of rock mass through discrete element method (e.g. Ding et al. 2014; Potyondy and Cundall 2004). A ratio of the smallest characteristic model length (i.e. the width of the specimen in this test) to median particle diameter (L/d_{50}) greater than 25 was suggested for the DEM simulation to obtain mechanical properties within a 2% variation level for rock modeling (Ding et al. 2014). In this study, L/d_{50} ratio of 27 was adopted to ensure the behaviour of rock mass is properly replicated. An illustration of the numerical model of intact rock specimen with the frictionless walls is shown in Fig. 1a.

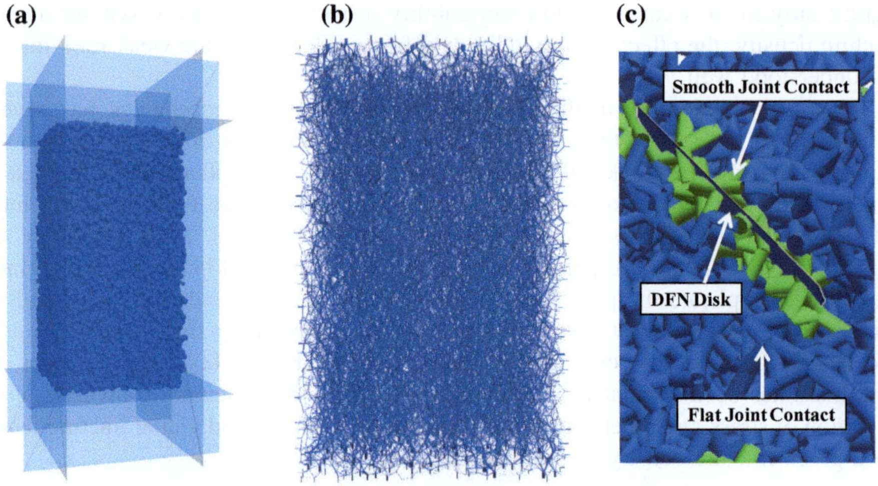

Fig. 1. **a** Numerical model of intact rock specimen under triaxial test with six rigid walls; **b** Contact force chain for intact rock specimen (line thickness and orientation represent force magnitude and direction, respectively), and **c** The smooth-joint contact model for simulating fractures

2.1.1 Flat-Joint Model (FJM) to Simulate Rock Particle Contacts

The flat-joint model (FJM) was adopted to simulate the particle contacts in the intact rock specimen. The FJM has several capabilities and advantages for simulating the behaviour of rock mass compared with the traditional linear parallel bond model. Figure 1b displays the contact force chain between particles. The flat-joint model allows partial breakage of the particle bonds as the contact between grains are discretized into several elements, and each element can be bonded or unbounded with friction. Comparing to the linear parallel bond model with a constant shear strength, the particle assembly bonded by the FJM follows the Coulomb criterion, which is controlled by cohesion, friction angle and normal stress.

2.1.2 Smooth-Joint Contact Model (SJM) to Simulate Rock Fractures

The smooth-joint contact model originally proposed by Pierce et al. (2007), can be assigned to the discrete fracture network (DFN) to replicate the characteristics of fractures in the rock mass. The discrete fracture network can be inserted with defined orientation, size, and position, and then the SJM can be assigned to contacts between particles. At the fracture contacts, the initial bonds are removed and smooth joints are defined in a direction parallel to the joint plane. The illustration of the SJM and DFN in the fractured rock model used in this study is shown in Fig. 1c.

2.2 Model Calibration for Intact Rock Specimen

The sample porosity has great influence on the behaviour of the rock mass in the DEM simulations (Zhang et al. 2014). The natural rock mass of the weak rock adopted in this study had a uniform porosity of 0.56 subjected to a confining pressure of 392 kPa (Sun et al. 2006). Therefore, in order to replicate the behaviour of the rock specimen, it was ensured that the numerical specimen had a porosity within 1% variation of the natural rock mass porosity after the isotropic compression loading. Then, a strain controlled loading was assigned to the specimen in the vertical direction similar to the loading done in the laboratory. The side walls were controlled by a wall-servo mechanism to maintain the cell pressure of 392 kPa throughout the test.

As the particle contacts in the numerical model were simulated using the flat-joint model adopting springs to transfer forces between particles, the strength of the specimen was controlled by the normal stiffness at each particle contact and the normal-to-shear stiffness ratio (k_n/k_s). The shear strength of the rock specimen was controlled by the cohesion (c_c) and friction coefficient (μ) between contacting particles. It should be noted that the Poisson's ratio of the specimen is mainly affected by the value of k_n/k_s. Through the trial and error procedures following the recommendation by Ding et al. (2014) and Potyondy and Cundall (2004), a number of numerical simulations of the rock specimen were conducted until the macro-properties could be properly reproduced. The micro-mechanical model parameters were calibrated resulting in a good agreement between the DEM prediction and laboratory measurements as observed in Figs. 2 and 3. The calibrated micro-mechanical parameters are summarised in Tables 1 and 2.

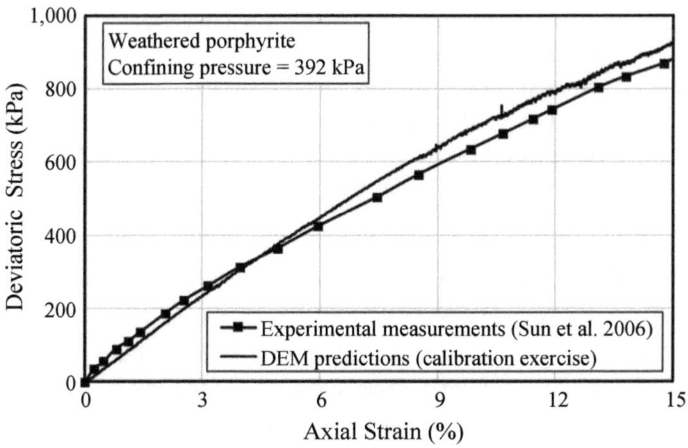

Fig. 2. Stress-strain curve of the intact rock specimen calibration exercise

2.3 Fractured Rock Simulation

For a controlled fracture density (ξ), which is defined as the fracture area over the volume of the specimen, various individual fracture lengths (η) were adopted to analyze the effects of the individual fracture length on the behaviour of the weak rock specimens. Referring to Prudencio and Jan (2007), the individual fracture length of 2.5 and 5 mm had been studied for non-persistent fracture distribution, which had the η/L

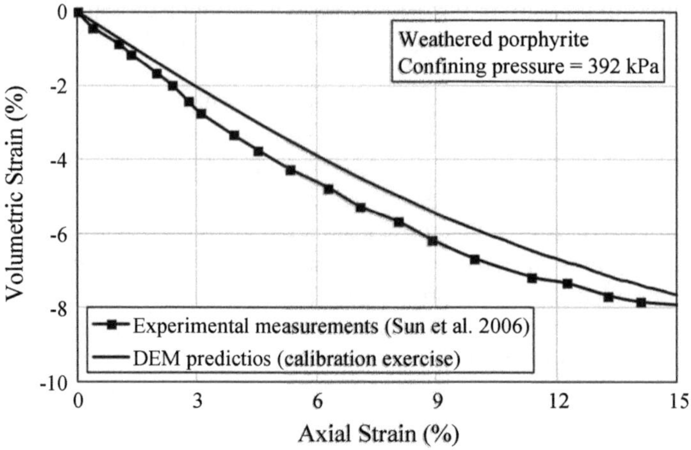

Fig. 3. Volumetric strain variations of the intact rock specimen during shearing—calibration exercise

Table 1. Summary of calibrated flat-joint parameters for the intact weak rock specimen

Parameter	Description	Value
d_{50}	Median grain diameter	3.5 mm
d_{max}/d_{min}	Ratio of maximum to minimum grain diameter	2
k_n	Contact normal stiffness	700 MPa/m
k_n/k_s	Normal-to-shear stiffness ratio of contacts	1.0
σ_c	Contact tensile strength	10 MPa
c_c	Contact cohesion	50 MPa
μ	Contact friction coefficient	0.5

Table 2. Summary of smooth-joint parameters for fractures

Parameter	Description	Value
k_{nsj}	Contact normal stiffness	400 MPa/m
k_{ssj}	Contact shear stiffness	100 MPa/m
μ_{sj}	Contact friction coefficient	0.7

of 0.05 and 0.1, respectively. However, the difference between predictions incorporating non-persistent fracture and continual fracture is still unclear. Therefore, in this study, the η/L value ranging from 0.028 to 1.4, covering the range from non-persistent fractures to one single continual fracture, has been studied. A controlled fracture density of $\xi = 7\ m^2/m^3$ was selected in this study. A fracture orientation of 45° was adopted since it could have great influence on the rock strength and deformability referring to Prudencio and Jan (2007) and Yang et al. (2008). The individual fracture lengths varying from 2.8 mm to 140 mm, adopted in this study are summarised in

Table 3. It should be noted that the tested specimens with shorter individual fracture length had plenty of small fractures distributed randomly over the rock mass, while for the individual fracture length of 140 mm, only one fracture was present in the specimen as illustrated in Fig. 4.

Table 3. Different fracture configurations used in the simulation of fractured rock sample

Specimen No.	Individual fracture length (η, mm)	Normalized fracture length (η/L)	Fracture density (ξ, m^2/m^3)	Confining pressure (kPa)
0	Intact	N/A	N/A	392
1	2.8	0.028	7	
2	5.6	0.056		
3	11.2	0.112		
4	28	0.28		
5	84	0.84		
6	140	1.4		

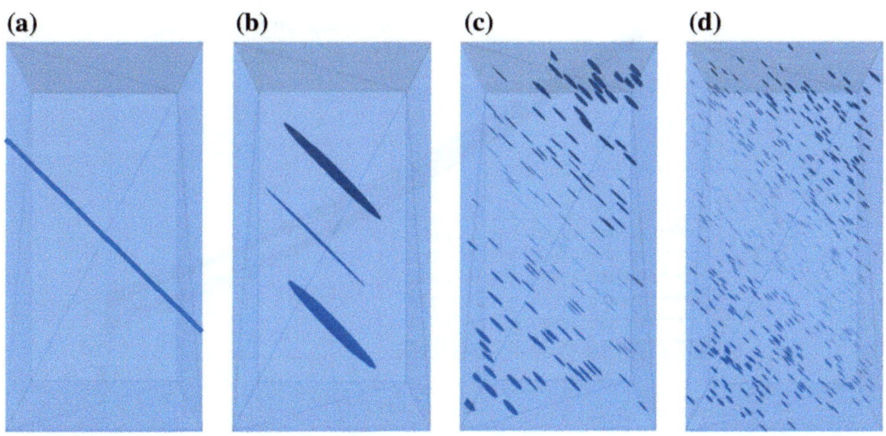

Fig. 4. Discrete fracture network for the rock specimen with individual fracture length of **a** 140 mm, **b** 84 mm, **c** 11.2 mm, and **d** 5.6 mm (*Note* the fracture density remains constant $\xi = 7\ m^2/m^3$)

3 Results and Discussion

The effects of the individual fracture length on the stress-strain behaviour and volume change characteristics of the fractured rock specimen are plotted in Figs. 5 and 6, respectively. Comparing Figs. 2 and 5, the pre-existing fractures contributed to a significant reduction in the strength and stiffness of the rock mass. As reported in Fig. 5, for a given fracture density (i.e. $\xi = 7\ m^2/m^3$), the reduction in strength varied

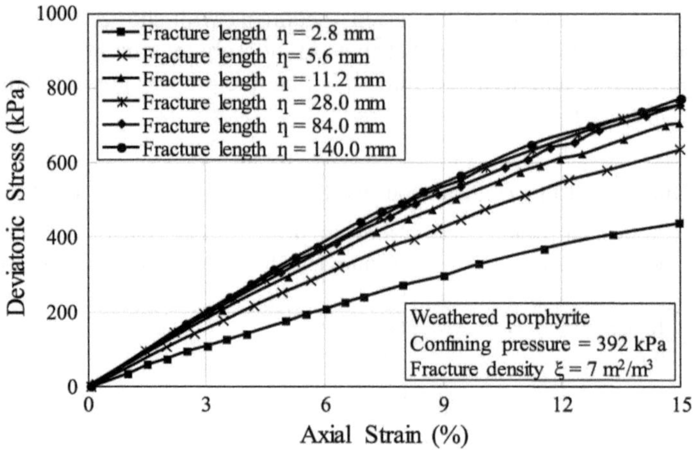

Fig. 5. Predicted stress-strain curves of fractured specimens with different individual fracture length

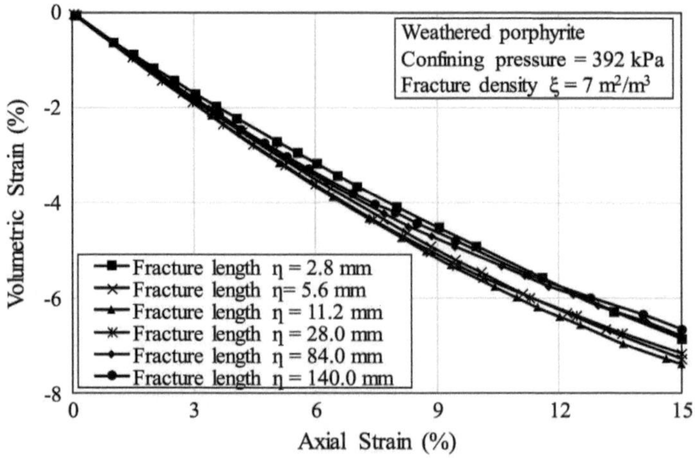

Fig. 6. Volumetric strain of fractured specimens with different individual fracture length

with the individual fracture length. It is noticed that plenty of non-persistent fractures with a short length (i.e. η/L less than 0.28) distributed over the specimen resulted in lower strength compared with the continual longer fractures. As expected, this observation is in a good agreement with Griffith's criterion (Bieniawski 1967), since new fissures are formed at the tip of the pre-existing non-persistent fractures when subjected to loading, generating a further reduction in the strength of the rock mass. However, as shown in Fig. 7, when the ratio of fracture length to the specimen width (i.e. the normalized fracture length, η/L) is greater than 0.28, the predicted strength remains unchanged. Indeed, for a given fracture density, the smaller the individual fracture length, the lower the strength and stiffness of the rock mass.

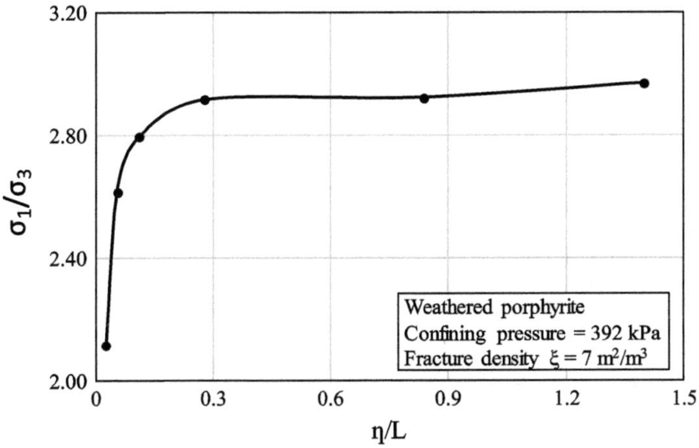

Fig. 7. Applied axial stress to cell pressure ratio (σ_1/σ_3) versus the ratio of individual fracture length to the specimen width (η/L) at 15% axial strain

As plotted in Figs. 3 and 6, the volumetric strains were less for the fractured rock specimens in comparison to corresponding values for the intact specimen. In this study, it has been assumed that fractures do not dilate (i.e. dilation angle = 0), hence, no volume change during shearing of fractures was expected. Therefore, as more fractures were introduced, the volume changes of the sample under applied shear loads decreased. Thus, as observed in Fig. 6, for a given fracture density (i.e. $\xi = 7\ m^2/m^3$), shorter but more abundant fractures could lead to further fractures during loading, resulting in less volume change of this contractive weak rock.

Indeed, wing fractures, which were connected to the tips of two neighboring fractures, tended to form easier when more abundant and therefore more closely spaced fractures (i.e. shorter fractures) were present. In addition, the total number of fracture tips were higher for the case of shorter but more abundant fractures, leading to further growth of wing fractures.

As shown in Fig. 8, at the same shear stress level, it is noticed that the specimens with shorter individual fractures (i.e. $\eta < 28$ mm) have experienced more significant volume contraction. As the weak intact rock adopted in this study had a high void ratio (i.e. e = 1.25), notable volume contraction had been observed during the shearing. Referring to Figs. 5 and 8, for a given deviatoric stress level, samples with shorter but more abundant fractures experienced larger shear strains, and therefore increased volume contraction. Referring to Fig. 8, for a given deviatoric stress (e.g. 400 or 500 kPa), the sample volume contraction was very sensitive to the normalized fracture length when $\eta/L < 0.112$, while for $\eta/L > 0.28$ the volume contraction prediction was almost insensitive to η/L ratio.

Figure 9 illustrates the grain displacement in the fractured specimens. It is noticed that for the individual fracture length of η = 140 mm or 84 mm, the grains were mainly moving on the fracture planes, while this trend became less apparent for shorter individual fractures since new tension fractures were formed around the pre-existing fractures. It should be noted that in this research visco-plastic effects as well as water

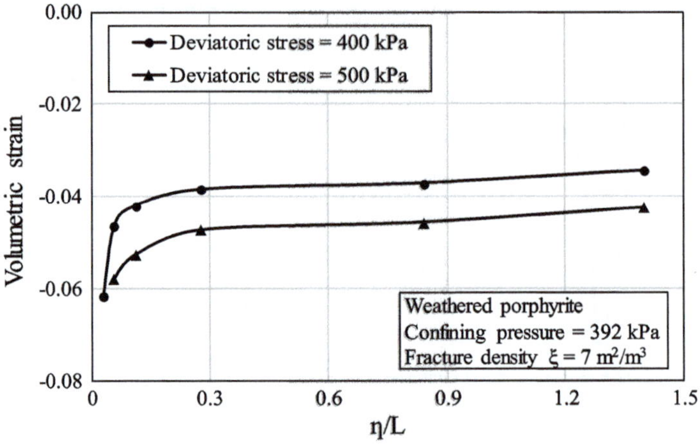

Fig. 8. Volumetric strain versus normalised fracture length for fractured rock specimens

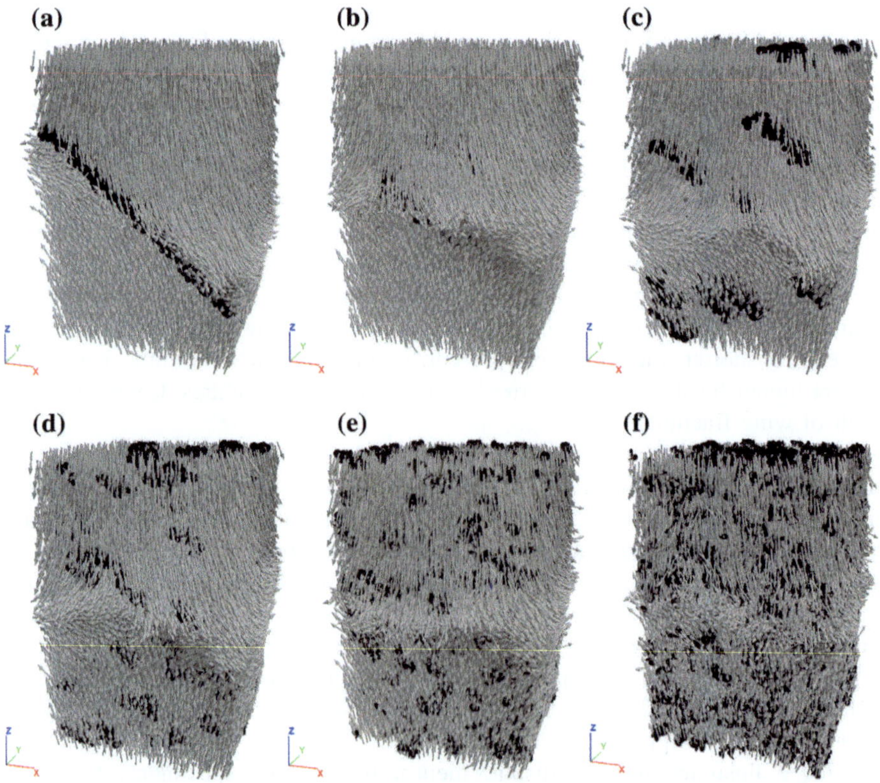

Fig. 9. Particle displacement representation for fractured specimens at the axial strain of 15% with normalised fracture length (η/L) of **a** 1.4, **b** 0.84, **c** 0.28, **d** 0.112, **e** 0.056, and **f** 0.028 (grey: particle displacement and black: smooth-joint model contacts for fractures)

effects have not been considered and therefore further research inspired by effects of geomaterial creep on deformation and excess pore water pressures can be conducted in future (see Le and Fatahi 2016; Le et al. 2017). In addition, for further research, authors recommend conducting stochastic analysis about effects of random distribution of fractures on predicted volume change and stress-strain responses of the fractured rocks.

4 Conclusions

In this study, for a given fracture density, the effects of the individual fracture length on the strength and deformability of rock mass has been investigated using the discrete element method. The flat-joint model has been adopted and calibrated to replicate the micro-parameters of the intact weak rock, while the smooth-joint contact model has been used to simulate the fractures. The stiffness and strength of fractured weak rock specimens were analyzed through triaxial tests with same confining pressure as the one used for calibration. The individual fracture length varied from one long single fracture to plenty of non-persistent short fractures. For a given fracture density of $\xi = 7\ m^2/m^3$, the shorter individual fractures resulted in more reduction in strength and stiffness of weak rock. When the fracture length to specimen width ratio $\eta/L > 0.28$, the effects of the individual fracture length were insignificant. This study sheds light on the understanding of the mechanical behaviour of fractured weak rock with different individual fracture lengths.

References

Bahaaddini, M., Sharrock, G., Hebblewhite, B.: Numerical investigation of the effect of joint geometrical parameters on the mechanical properties of a non-persistent jointed rock mass under uniaxial compression. Comput. Geotech. **49**, 206–225 (2013)

Bieniawski, Z.T.: Mechanism of brittle fracture of rock: part I—theory of the fracture process. Int. J. Rock Mech. Mining Sciences & Geomechanics Abstracts, vol. 4, Elsevier, pp. 395IN11405-11404IN12406 (1967)

Chiu, C.-C., Wang, T.-T., Weng, M.-C., Huang, T.-H.: Modeling the anisotropic behavior of jointed rock mass using a modified smooth-joint model. Int. J. Rock Mech. Min. Sci. **62**, 14–22 (2013)

Ding, X., Zhang, L., Zhu, H., Zhang, Q.: Effect of model scale and particle size distribution on PFC3D simulation results. Rock Mech. Rock Eng. **47**(6), 2139–2156 (2014)

Goodman, R.E., Ahlgren, C.S.: Evaluating safety of concrete gravity dam on weak rock: Scott Dam. J. Geotechnical Geoenvironmental Eng. **126**(5), 429–442 (2000)

Klein, S.: An approach to the classification of weak rock for tunnel projects. In: Proceedings of the Rapid Excavation and Tunneling Conference, pp. 793–806 (2001)

Le, T.M., Fatahi, B.: Trust-region reflective optimisation to obtain soil visco-plastic properties. Eng. Comput. **33**(2), 410–442 (2016)

Le, T.M., Fatahi, B., Khabbaz, H., Sun, W.: Numerical optimization applying trust-region reflective least squares algorithm with constraints to optimize the non-linear creep parameters of soft soil. Appl. Math. Model. **41**, 236–256 (2017)

Nguyen, L., Fatahi, B.: Behaviour of clay treated with cement and fibre while capturing cementation degradation and fibre failure–C3F Model. Int. J. Plast **81**, 168–195 (2016)

Nguyen, L., Fatahi, B., Khabbaz, H.: A novel model to simulate the behaviour of cement-treated clay under compression and shear. In Geo-China **2016**, 152–158 (2016)

Nguyen, L., Fatahi, B., Khabbaz, H.: Development of a constitutive model to predict the behavior of cement-treated clay during cementation degradation: C3 model. Int. J. Geomech. **17**(7), 04017010 (2017)

Nickmann, M., Spaun, G., Thuro, K.: Engineering geological classification of weak rocks. In: Proceedings of the 10th International IAEG Congress (2006)

Pierce, M., Cundall, P., Potyondy, D., Mas Ivars, D.: A synthetic rock mass model for jointed rock rock mechanics: meeting Society's challenges and demands. In: 1st Canada-US Rock Mechanics Symposium, Vancouver, vol. 1, pp. 341–349 (2007)

Potyondy, D., Cundall, P.: A bonded-particle model for rock. Int. J. Rock Mech. Min. Sci. **41**(8), 1329–1364 (2004)

Prudencio, M., Jan, M.V.S.: Strength and failure modes of rock mass models with non-persistent joints. Int. J. Rock Mech. Mining Sci. **44**(6), 890–902 (2007)

Santi, P.M.: Field methods for characterizing weak rock for engineering. Environ. Eng. Geosci. **12**(1), 1–11 (2006)

Sun, D.A., Feng, T., Matsuoka, H.: Stress–strain behaviour of weathered weak rock in middle-sized triaxial tests. Can. Geotech. J. **43**(10), 1096–1104 (2006)

Tsoutrelis, C., Exadaktylos, G.: Effect of rock discontinuities on certain rock strength and fracture energy parameters under uniaxial compression. Geotech. Geol. Eng. **11**(2), 81–105 (1993)

Yang, S., Jiang, Y., Xu, W., Chen, X.: Experimental investigation on strength and failure behavior of pre-cracked marble under conventional triaxial compression. Int. J. Solids Struct. **45**(17), 4796–4819 (2008)

Numerical Analysis of the Long-Term Performance of Energy Piles in Sand

Kang Fei(✉), Wei Hong, and Jian Qian

Institute of Geotechnical Engineering,
Yangzhou University, Yangzhou 225127, China
kfei@yzu.edu.cn

Abstract. Energy piles are piles equipped with heat exchange pipes through which a heat-carrying fluid circulates and exchanges heat with the ground. This technology couples the structural role of classical pile foundations with the energy supply of heat exchangers. During heating and cooling processes, the temperature of the energy pile and the ground will change seasonally. Due to the thermal displacement incompatibility between the pile and the soil, the load transfer mechanism of energy piles is different to that of conventional piles which are only subjected to mechanical loadings. In order to improve the understanding of the long-term performance of energy piles in sands, a series of coupled thermal-stress finite element analyses were carried out. In the analyses, the bounding surface plasticity model was used to describe the nonlinear behavior of sands under monotonic and cyclic loadings. The thermally induced displacement and axial force in the pile, the thermally induced change in the soil stress, and the ultimate pile resistance after thermal cycles were discussed. The numerical results indicated that the soils around the energy pile were subjected to cyclic mechanical loadings caused by repeated temperature variations. The accumulation of plastic strains resulted in a significant increase in the pile head settlement for the free head pile and a significant decrease in the pile head reaction force for the restrained head pile. During the reloading stage, the thermally induced decrease in the shaft resistance was compensated by the soil dilatancy, the ultimate pile resistance after thermal cycles did not change remarkably.

Keywords: Energy pile · Long-term performance · Thermal cycles
Numerical analysis · Bounding surface plasticity model

1 Introduction

Ground source heat pumps (GSHPs) are one of the most efficient technologies for the geothermal energy utilization. In a conventional GSHP system, the heat exchange is accomplished by circulating a working fluid through the heat transfer pipes installed in bore holes. In order to reduce the installation costs of GSHP systems, the heat transfer pipes may be incorporated into conventional foundation piles. Such piles equipped with fluid carrying pipes are usually called as energy piles or heat exchanger piles (Brandl 2006; Laloui et al. 2006; Adam and Markiewicz 2009). During the heating/cooling operation of a GSHP system, energy piles will be subjected to seasonal temperature

W. Steyn et al. (eds.), *Pavement Materials and Associated Geotechnical Aspects of Civil Infrastructures*, Sustainable Civil Infrastructures,
https://doi.org/10.1007/978-3-319-95759-3_5

variations. The effects of thermal loads on the mechanical behavior of energy piles have attracted the attention of many researchers. Based on the field data measured by Brandl (2006), Laloui et al. (2006), and Bourne-Webb et al. (2009), the maximum thermally induced stresses in the pile were about 100–300 kPa/°C, the exact values mainly depended on the soil and the pile end-restraint conditions. Since the temperature change of energy piles is generally from −15 to +20 °C, the risk of structural damages due to the temperature variation is relatively small (Di Donna and Laloui 2015). The effects of the temperature variation on the pile capacity and the pile settlement have become the focus of research.

According to the field test of an energy pile in dense sand, Wang et al. (2015) reported that the shaft resistance increased at least 14% when the pile temperature increased by about 20 °C. After cooled naturally, the ultimate shaft resistance returned to its initial value, no decrease in the shaft resistance was observed after the heating and recovery cycle. Ng et al. (2015) carried out a series of centrifuge tests on energy piles in medium dense saturated sand. They also found that the pile capacity increased with the pile temperature. This trend is believed to be the result of the increased horizontal stress caused by the radial thermal expansion of the pile (McCartney and Rosenberg 2011). However, Goode et al. (2014) observed that there was no apparent change in the pile capacity based on the centrifuge experiments with dry sand. Olgun et al. (2014) pointed out that the contribution of the radial expansion to the shaft resistance was not as high as expected, the change of the moisture content and the soil mechanical behavior may also have important effects.

It is worth noting that the energy pile will experience seasonally expansion and contraction during the entire service period. As a result, the surrounding soil will be subjected to cyclic loadings, which may have an important impact on the bearing capacity and the settlement of the energy pile, especially for the floating or the semi-floating pile (Suryatriyastuti et al. 2014; Ng et al. 2016). Ng et al. (2014a, b) performed centrifuge tests on floating energy piles in clay. They observed that for the piles in lightly and heavily overconsolidated clay, the cumulative pile settlements after five thermal cycles were 3.8 and 2.1% pile diameter respectively, which were significantly larger than those caused by the static working load. Kalantidou et al. (2012) also studied the effects of temperature cycles on the pile settlement. It was shown that when the mechanical load was high, the irreversible pile settlement after thermal cycles could not be ignored. A similar conclusion has been made by Yavari et al. (2014) using laboratory scale tests. One possible reason for the irreversible settlement is that the thermal deformation of the energy pile may modify the stress state of the surrounding soil and induce the plastic strain.

Saggy and Chakrabotry (2015) investigated the effects of thermal cycles on the pile capacity and the displacement by finite element analyses. In their study, the sand behavior was described by the Mohr-coulomb model. They suggested that the thermal loadings increased the limit shaft resistance for the pile embedded in dense sand, while it has no effect for the pile in loose sand. This phenomenon occurred because the dense sand provided a stronger constraint to the radial expansion of the pile. However, this conclusion may be questioned since the strain softening of the soil and the accumulation of plastic strain during thermal cycles were not modeled reasonably.

At present, the effects of thermal cycles on the changes in the stress state of the soil and the long-term performance of energy piles are not well understood yet. In order to improve the understanding of the long-term performance of energy piles in sand, a series of coupled thermal-stress finite element analyses has been carried out. Based on the results, the mechanical behavior of energy piles during and after thermal cycles was analyzed.

2 Numerical Model Description

2.1 Numerical Analysis Plan

In this study, the long-term performance of single concrete energy piles in both loose and dense dry sand was investigated by ABAQUS software. The modeled piles were 1.0 m in diameter and 20.0 m in length.

For each pile, three sets of analyses have been carried out. In the first set of analyses, the pile was static loaded to failure at a constant temperature. Based on the predicted load-settlement relationship, the ultimate pile resistance was determined.

For the second set of analyses, thermal cycles were applied at the working load level (with a factor of safety of 2.5). Two types of pile head fixities, i.e. a free head pile and a restrained head pile, were considered. The restrained pile considered in this study was an approximation to the single energy pile within a pile group consists of energy and conventional piles. At this situation, the thermally induced displacement at the pile head will be constrained by the raft and can be regarded as zero.

To evaluate the influence of the past thermal cycles on the ultimate pile resistance, the pile was reloaded to failure in the third set of analyses.

2.2 Material Models and Model Parameters

The pile was considered to be thermo-elastic and was modeled as a linear elastic material. The relevant parameters were shown in Table 1.

Table 1. Parameters for the pile

Density ρ kg/m^3	Young's modulus E/GPa	Possion's ratio v	Linear coefficient of thermal expansion $\alpha/10^{-6}$/°C	Thermal conductivity λ/W/(m · °C)	Specific heat c_p/J/(kg · °C)
2500	30	0.18	10	2.0	970

For the dry sands considered in the study, the thermal characteristics were evaluated taking into account the volumetric fractions of the solid soil particles and the pore air. The adopted thermal constants listed in Table 2 were estimated based on the typical values for sand particles

As reported by Yavari et al. (2016), the compressibility and the shear strength of sand was independent of the temperature, so the thermal effects on the mechanical

Table 2. Parameters for the sands

Parameter		Loose Sacramento River sand	Dense Sacramento River sand
Linear coefficient of thermal expansion	$\alpha/10^{-6}/°C$	5.3	6.2
Specific heat	c_p/J/(kg · °C)	946	938
Thermal conductivity	λ/W/(m · °C)	1.34	1.55
Dry density	ρ_d/kg/m^3	1.45	1.67
Elastic swelling modulus	κ	0.0075	0.01
Poisson's ratio	ν	0.2	0.2
Critical void ratio at 100 kPa	Γ	0.84	0.97
Slope of critical state line in e–ln p plane	λ	0.076	0.076
Ellipse aspect ratio	ρ	1.65	2.2
Residual friction angle in compression	φ_c/°	36.2	34.6
Residual friction angle in tension	φ_e/°	36.2	34.6
Peak friction angle	φ_p/°	43.8	35.7
Plastic modulus	h_0	2	2

behavior of sands were neglected in the study. To describe the strain softening/hardening, the stress dilatancy and the accumulation of irreversible strains during cyclic loadings, the bounding surface model proposed by Bardet (1986) was used. This model has nine parameters which can be determined from the results of conventional triaxial tests. The values of material constants for loose and dense Sacramento River sand (Bardet 1986) were used in the analyses (Table 2).

The soil-pile interface was modeled by a layer of thin elements, and the constitutive parameters were taken as the same as those for the surrounding sand.

2.3 Initial and Boundary Conditions

The numerical analyses were carried out in axisymmetric conditions. As shown in Fig. 1, the height and the width of the model domain were taken to be 2L and L respectively, where L is the pile length. Both vertical and horizontal displacements were fixed on the bottom of the model domain, and only horizontal displacements were constrained on the lateral sides.

The initial stresses due to gravity were calculated by the bulk unit weights and a coefficient of later earth pressure $K_0 = 0.5$. It should be noted that the pile was considered to be a bored pile, so the installation effects were not considered in the analyses.

The initial temperature of the pile and the ground was 15 °C. To model the thermal operation process of the energy pile, the pile temperature was set to change cyclically.

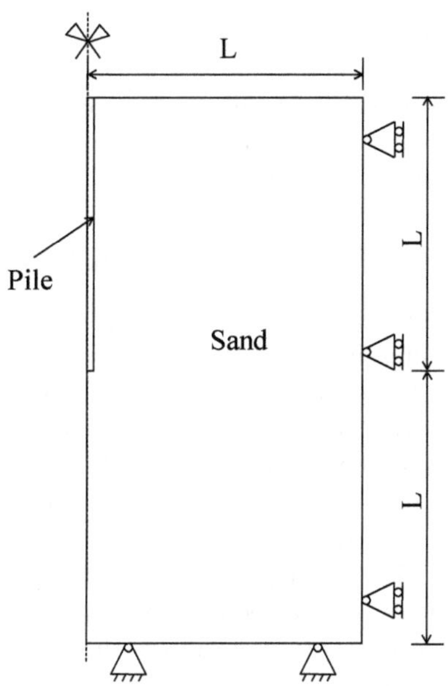

Fig. 1. Sketch of the model

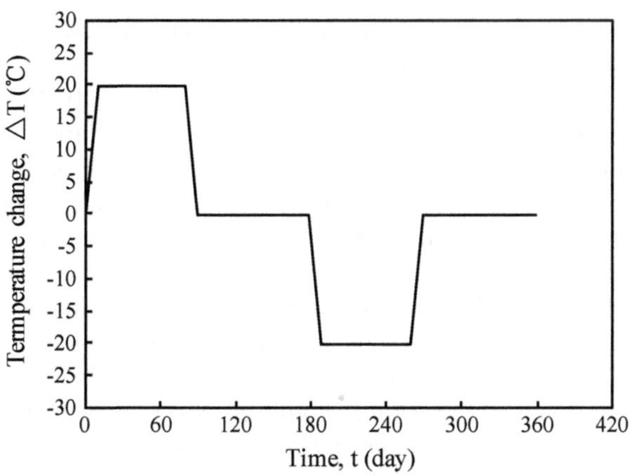

Fig. 2. Temperature variation of pile in one cycle

The temperature variation curve in one cycle is shown in Fig. 2. During the first 90 days (summer mode), the pile temperature was increased by 20 °C in 10 days and then kept constant during the rest time. At the end of this period, the pile temperature was decreased to the initial value and then kept constant for another 90 days. During the following winter mode, the opposite temperature change was applied. In the study, a total of 20 cycles of thermal loading was applied to study the long-term performance.

3 Numerical Results

In this section, the pile response during thermal cycles is analyzed for different pile head fixities. The ultimate pile resistance, the pile displacement, and the pile axial force for all the cases are summarized in Table 3.

3.1 Ultimate Resistances of Piles Without Thermal Cycles

Figure 3 shows the predicted load-settlement relationships for the pile in the loose and dense sand. The results are presented in terms of total resistance Q_t, the shaft resistance Q_s, and the base resistance Q_b. The shaft resistance was mobilized much earlier than the base resistance. The base resistance did not reach a limiting value even at a displacement of 0.1 m. Based on the point where the slope of the load settlement curve significantly decreased, the ultimate pile resistances$Q_u$were 5596 kN (loose sand) and 35,342 kN (dense sand). The corresponding mobilized base resistances were 648kN (Q_b/Q_u = 11.6%) and 6185 kN (Q_b/Q_u = 17.5%) respectively. Since the load was mainly carried by the shaft resistance, both piles were friction piles. By applying a factor of safety of 2.5, the design working load was chosen as 2238 kN and 14,137 kN for the pile in the loose and dense sand respectively.

3.2 Thermally Induced Pile Displacement

3.2.1 Free Head Pile

The displacement distributions of the free head pile after n thermal cycles are shown in Fig. 4. In the figure, downward displacements are taken positive. The letter 1-H and 1-C represent the heating (ΔT = 20 °C) and cooling (ΔT = −20 °C) stage of the first thermal cycle respectively. As expected, the pile expanded during heating and contracted during cooling. For the pile in the loose sand, the pile head heaved by 2.8 mm after the first heating. At the end of the following cooling period, the pile head was 3.2 mm lower than the initial position. Due to the plastic soil strains developed below the pile tip, an irreversible settlement was observed at the pile head when the first thermal cycle finished. The pile head settlement increased gradually to 13.1 mm after 20 cycles, which was about 2.4 times the initial static settlement caused by the mechanical load. The potential influence of the accumulated settlement on the serviceability of the pile and the structure should be considered in the design. For the energy pile in the dense sand, a similar displacement pattern was observed in Fig. 4b. Because the dense sand has higher ability to resist cyclic loading, the increment of the pile settlement was relatively smaller than that for the loose sand. The final settlement was only 1.4 times the initial settlement.

The location of the null point, characterized by zero thermally induced pile displacement, can be identified by comparing the pile displacement distributions at different stages. For the free head pile in the loose sand, the null point of the first heating phase was at a depth of 14.2 m (indicated as NP-H1 in Fig. 4), which was deeper than the midpoint of pile. This phenomenon may be attributed to the increasing soil resistance with the depth. During the subsequent cooling period, the deeper part of the pile-soil interface underwent unloading while the upper part underwent loading. Due to

Table 3. Summary of numerical results

Pile head condition	Soil type	Ultimate pile resistance/kN		Working load/k*N*	Pile head settlement/mm		Maximum change of pile axial force/kN	
		Without thermal cycles	With thermal cycles		Before thermal cycles	After thermal cycles	Compression	Tension
Free	Loose sand	5596	4850	2238	5.5	13.1	591	290
	Dense sand	35,342	34,400	14,137	29.8	42.6	317	384
Restrained	Loose sand	5596	4700	2238	5.5	5.5	1170	−6239
	Dense sand	35,342	34,280	14,137	29.8	29.8	1299	−9064

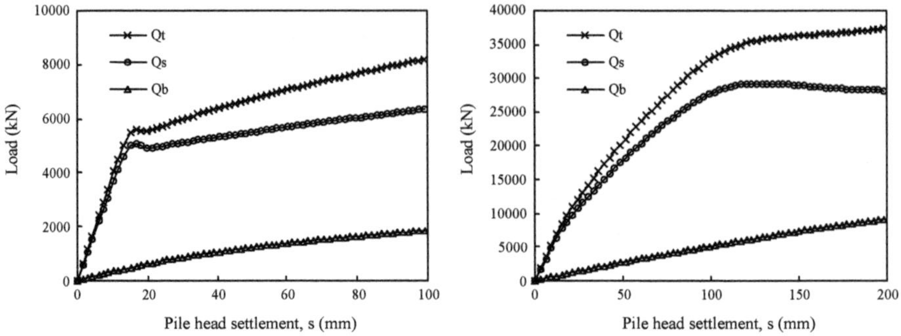

(a) Pile in loose sand (b) Pile in dense sand

Fig. 3. Load-settlement relationships

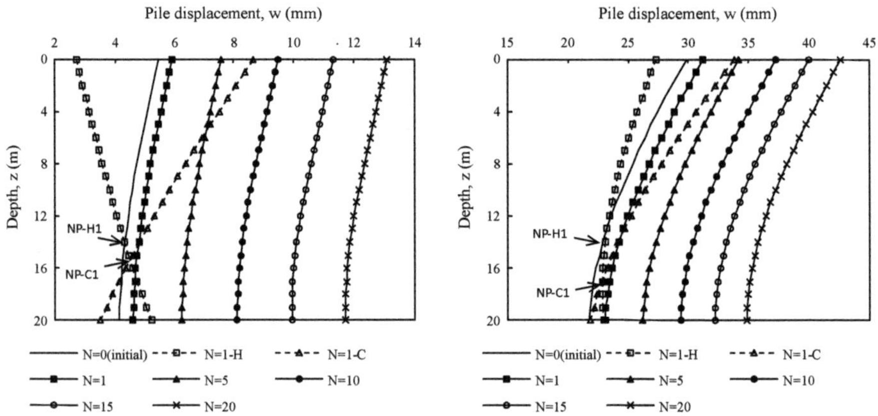

(a) Pile in loose sand (b) Pile in dense sand

Fig. 4. Displacement distribution of free head pile

the difference between the loading and unloading stiffness, the pile's contraction mobilized greater resistance over the lower section, and thus the null point of the cooling (NP-C1) was deeper that of the heating phase, at around 15.8 m depth. During the subsequent cycles, the locations of the null points did not change significantly. For the pile in the dense sand, the location of the null point varied from 13.2 m (heating) to 16.8 m (cooling).

3.2.2 Restrained Head Pile

For the restrained pile, the pile head was fixed at the beginning of the thermal cycles, so the whole part of the pile displaced in the same direction when heated or cooled (Fig. 5). In other words, the null point of the restrained pile was always at the pile head.

Another important finding is that the displacement at the pile tip increased with the number of thermal cycles, which indicated that the restrained head pile has a tendency

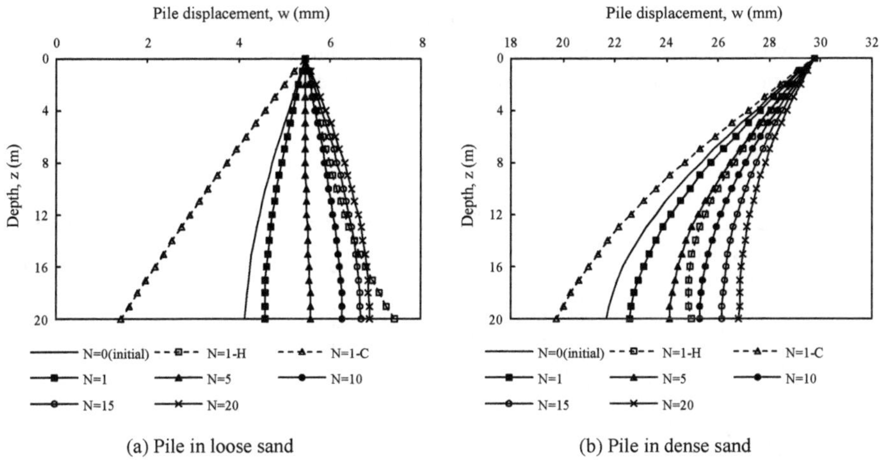

Fig. 5. Displacement distribution of restrained head pile

to elongate during thermal cycles. The main reason for this phenomenon is that the soil experienced significant plastic strains induced by the cyclic loading, so the soil settled more than the pile. Therefore, a downdrag load (or negative shear stress) was induced along the interface of the pile and the soil. Because the loose sand was easier to deform, the thermal cycles resulted in a more significant increase in the axial elongation of the pile. At the end of the thermal operation, the pile in the loose sand was in tension for the entire length.

3.3 Thermally Induced Axial Force in Pile

3.3.1 Free Head Pile

The profiles of the axial force P at different times were shown in Figs. 6 and 7. It is obvious that the effects of thermal cycles on the axial forces depended on the restraint condition and the ground type. For the free head pile in the loose sand, the axial force distribution during the first thermal cycle agreed well with that observed by Bourne-Webb et al. (2009) for the pile which was unrestrained at the pile top and partially restrained at the pile toe. The maximum changes occurred slightly below the mid-depth point. The increment in the axial force was about 280 kN (20% of the initial value at the same depth) for eating and −290 kN (26%) for cooling. After one thermal cycle, a small increase and decrease of the axial force was observed for the upper and the lower part of the pile, respectively. The change of the axial force was observed to increase with the number of thermal cycles. Due to the accumulation of the residual stress, the long-term distribution of the axial load was quite different to the initial shape. After the end of the thermal operation, the axial load first increased and then decreased with the depth. The arising of this phenomenon can be related to the relative displacement between the pile and the soil. Along the lower part of the pile, the pile section moved downward due to the deformation of the soil under the pile base, so the skin force was increased. While at the upper part of the soil, the negative skin force was generated due to the settlement of the soil caused by stress reversals. For the free head pile in the dense sand (Fig. 6b), the variation of the axial force was relative small due to the large static load and the high stiffness of the dense sand.

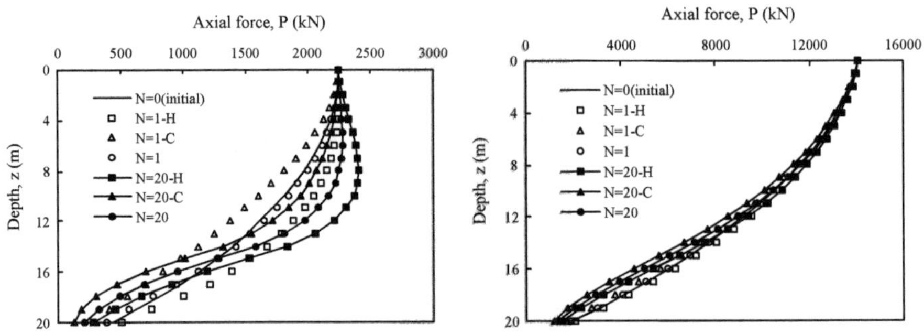

(a) Pile in loose sand (b) Pile in dense sand

Fig. 6. Axial force distribution of free head pile

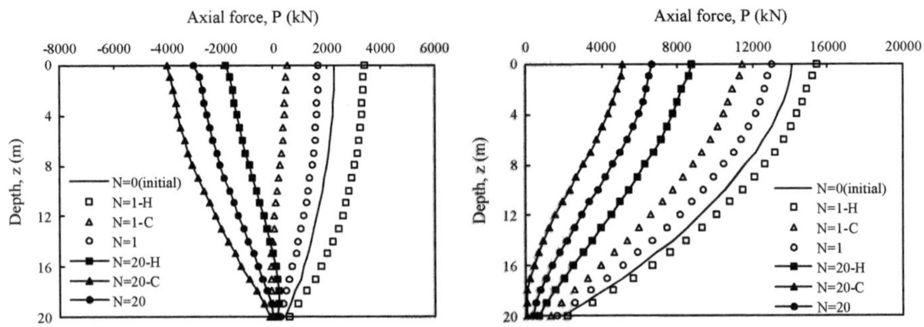

(a) Pile in loose sand (b) Pile in dense sand

Fig. 7. Axial force distribution of restrained head pile

3.3.2 Restrained Head Pile

If the pile head was fixed during thermal cycles, the compressive axial forces were found to decrease with cycles for both piles in the loose and dense sand. Corresponding to the displacement results in Fig. 5, the reduction in the axial force was more apparent for the pile in the loose sand. In this case, the maximum tensile force developed at the pile head was 4010 kN (5105 kPa). The risk of tensile failure should be considered in the design. It is also worth mention that in a pile group, the reduction of the pile head reaction will generate an additional load on the neighboring pile. The redistribution of stresses should also be taken into account.

3.4 Thermally Induced Change in Soil Stress

3.4.1 Free Head Pile

The stress paths of the soil elements adjacent to the pile shaft at 8, 15, and 18 m depths during and after thermal cycles are shown in Fig. 8. The results of the pile without thermal cycles are also given in the figure for reference. The corresponding shear stress at the pile-soil interface is plotted against the shear strain in Fig. 9. During the axial loading of the pile in the loose sand, the soil elements at depth depths (z = 15 and 18 m) were observed to be contractive and the mean stress decreased with the increase of the deviatoric stress. After the phase transformation state was passed, the soil dilated slightly and the stress path was inverted from left to right to reach the critical state. At shallower depth (z = 8 m), the soil was easy to dilate since the confining stresses were relatively small. Because the tendency to dilate was constrained by the pile shaft, the horizontal stress and the mobilized shear stress increased gradually. For the case of dense sand, the effect of dilation was more obvious. The shear stresses developed continuously while the stress states moved along the failure surface until constant soil volumes were reached at large shear strains.

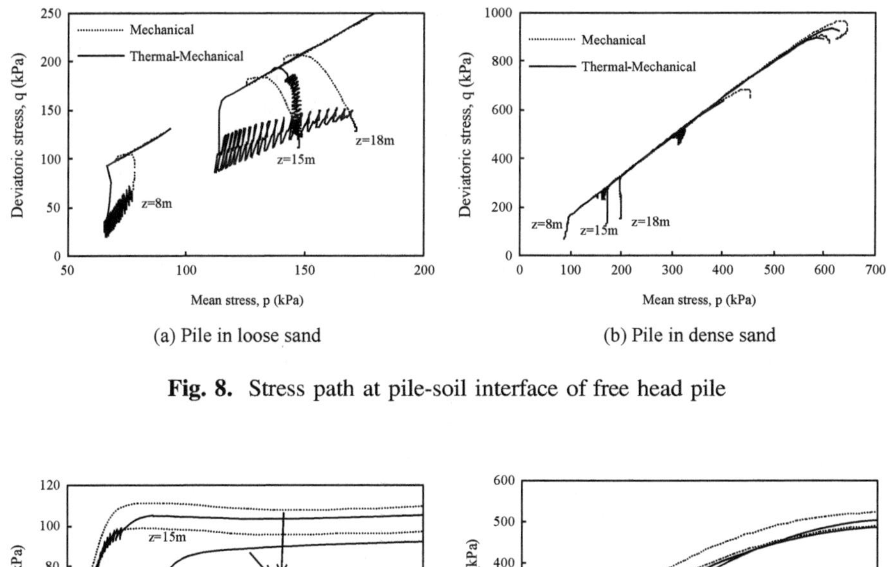

Fig. 8. Stress path at pile-soil interface of free head pile

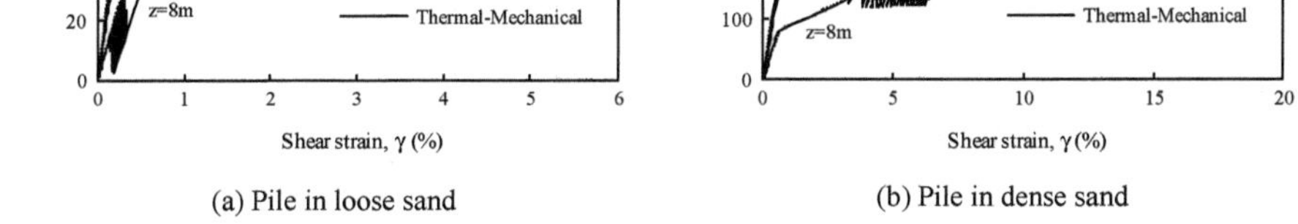

Fig. 9. Development of shear stress at pile-soil interface of free head pile

During thermal cycles, different patterns of stress responses were observed for the soils at different depths. This occurrence can be associated to the specific displacement mode developed at these locations. Below the null point, the thermally induced expansion pushed down the pile and led to an increase in the shear stress. For the upper part of the pile, the upward expansive deformation decreased the relative pile-soil displacement and the shear stress. During the process of cooling, the opposite variation of the shear stress would occur. As expected, the shear stresses of the soils at z = 8 and z = 18 m changed cyclically. However, because the null point of cooling was deeper than the null point of heating, the shear deformation of the soil between these two points was always increased during the heating/cooling cycles. Hence, the soil at z = 15 m was sheared monotonically and the response was similar to that of the pile subjected to mechanical loading case.

Because the amplitude of the pile expansion/contraction was mainly controlled by the temperature change and could be regarded as constant during thermal cycles. For the soils at z = 8 and 18 m, the soils were cyclic loaded in the stain controlled condition. Due to the accumulation of plastic strains in each cycle, the mean stresses and the shear stresses at the pile-soil interface decreased with the number of thermal cycles. The reduction of stresses was more apparent for the pile in the loose sand. This tendency of progressive reduction in the stress was consistent with that observed during the dynamic installation of piles. As a result, the stress paths moved progressively to the bottom left. For the soil at z = 15 m, because the soil was nearly sheared in a monotonic way during thermal cycles, the trajectory of the stress path was close to that under the static load. The increase of the shear stress near this point compensated for the friction fatigue at other depths, so the force equilibrium of the pile was maintained.

3.4.2 Restrained Head Pile

In this case, because the null point of the heating and cooling was fixed at the pile head, the shear stresses along the entire length of the pile increased simultaneously during heating and decreased simultaneously during cooling. Because the response at z = 15 m was similar to that at z = 18 m, only the results of the soil at z = 8 and 18 m are given in Figs. 10 and 11. As shown, the reductions of the mean stresses and the shear stresses were also evident. Because the thermal induced displacement of the pile increased with the distance to the null point, the variations of soil stresses were more apparent at deeper locations. For example, the reduction of the mean stress at z = 8 and 18 m were 17.3% and 36.3% for the pile in the loose sand, and were 7.2% and 43.0% in the case of dense sand. Due to the same reason, the decrease in the shear stress at z = 18 m was larger than that in the free head pile case. For the pile in loose sand, a negative shear stress was observed before the reloading started.

3.5 Ultimate Pile Resistance After Thermal Cycles

3.5.1 Free Head Pile

As shown in Fig. 9, for the free head pile in loose sand, the ultimate shear resistance mobilized at z = 8 m was 51.3 kPa, which was only 2.7 kPa less than that in the case of pile without thermal cycles. It implies that the stress reduction at this location after thermal cycles was almost fully compensated by the soil dilatancy during the reloading

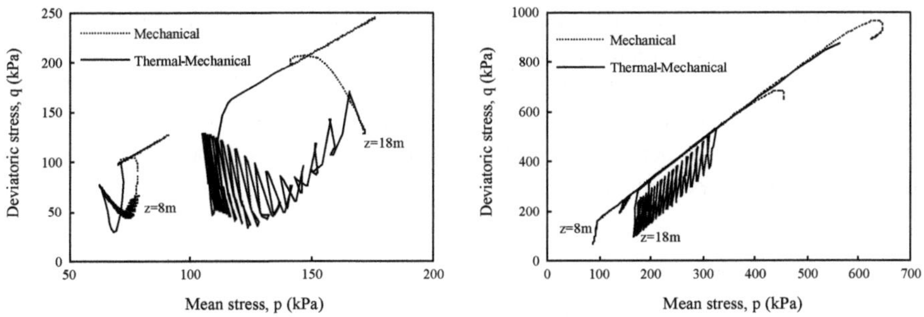

(a) Pile in loose sand (b) Pile in dense sand

Fig. 10. Stress path at pile-soil interface of restrained head pile

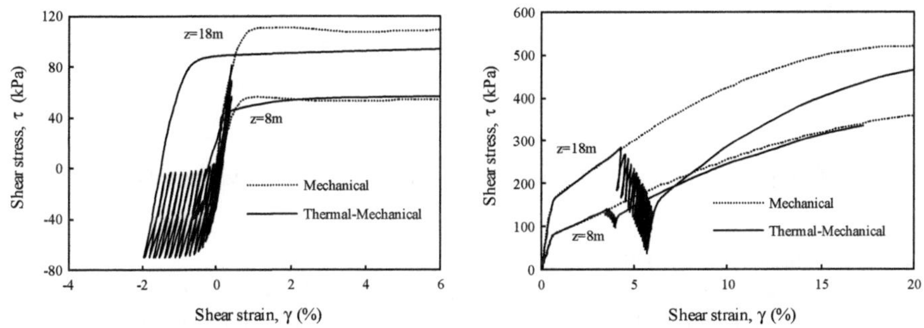

(a) Pile in loose sand (b) Pile in dense sand

Fig. 11. Development of shear stress at pile-soil interface of free head pile

stage. A different tendency was observed for the soil at z = 18 m. With the soil parameters applied and the specific stress condition, the soil was contractive in the subsequent reloading. The change of the soil state after thermal cycles decreased the ultimate shear resistance form 115.3 kPa to 88.2 kPa. For the soil at z = 15 m, the ultimate shear resistance was increased from 97.2 kPa to 106.9 kPa due to the increased horizontal stress after thermal cycles. Similar findings can also be found for the pile in dense sand.

To investigate the effect of thermal cycles on the long-term capacity, the load settlement curves with and without the thermal cycles are compared in Fig. 12. Since the load was mainly undertaken by the shaft resistance, the densification of the soil beneath the pile toe contributed little to the ultimate pile resistance and the load-settlement relationship. As mentioned before, the stress reduction maybe fully or partially compensated by the soil dilatancy, hence the magnitude of the total resistance did not change much. After 20 thermal cycles, the decreases in the total ultimate resistance were 13 and 3% for the free head pile in the loose and dense sand. It should be noted that due to the degradation effect at the pile-soil interface and the change in the soil state, the pile displacement required to mobilize ultimate resistance increased after thermal cycles. The overall stiffness of the pile-soil system decreased slightly after thermal cycles. Also taking into account the accumulated settlement during the thermal operation, the long-term performance of the energy pile seems to be mainly controlled by the settlement rather than the capacity.

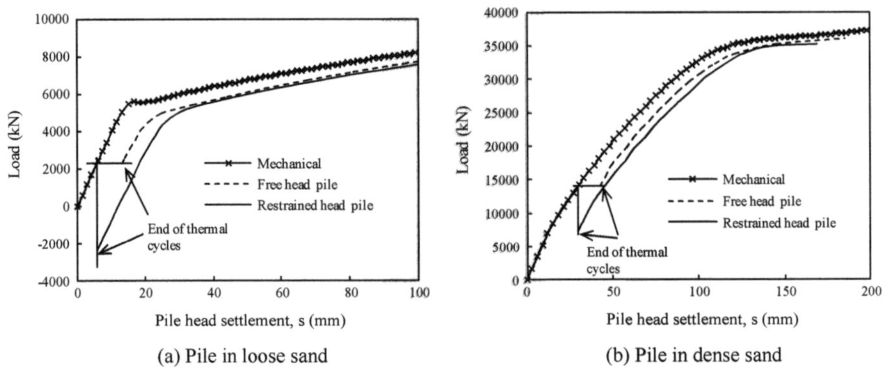

Fig. 12. Comparison of load settlement curves of piles with and without thermal cycles

3.5.2 Restrained Head Pile

The load settlement curves of restrained head piles in the loose and dense sand are also plotted in Fig. 12. Though the negative shear stress may be induced by the application of thermal cycles, the pile will move down relative to the soil in the subsequent reloading stage, and the negative shear stress will be eliminated. Similar to that observed for free head piles, the reductions of the horizontal stress and the mobilized shear stress recovered to a large extent in the subsequent reloading. The total resistances were reduced by 16% and 3% for the piles in loose and dense sand.

4 Conclusion

To improve the understanding of the effects of thermal cycles on the long-term performance of the single energy pile, coupled thermal-stress analyses were carried out. Based on the results, the following conclusions may be drawn:

(1) Due to the irreversible soil strains caused by cyclic loadings, the pile head settlement for the free head pile increased with the number of thermal cycles. After 20 cycles, the pile head settlement was 2.4 and 1.4 times the initial static settlement for the pile in the loose and dense sand, respectively. For the restrained head pile, a significant decrease in the pile head reaction was observed. For this case, the risk of tensile failure should be considered.

(2) The free head pile tended to have different development patterns of the shear stresses at different depths. For the soils located between the null point of heating

and the null point of cooling, the shear stresses developed monotonically despite the cyclic loading condition, while the reductions of the horizontal stress and the shear stress were observed for the other parts of the pile. In the case of the restrained head pile, the stress reduction was observed along the entire length of the pile.

(3) The reduction of the shaft resistance was recovered remarkably due to the soil dilatancy in the subsequent reloading. Comparing with piles subjected to mechanical loads, the ultimate pile resistance was reduced in the range of 13% to 16% for piles in the loose sand, and was reduced by 3% for those in the dense sand. Therefore, for the long-term design of energy piles in sand, the main problem is the settlement rather than the bearing capacity.

Acknowledgements. The authors acknowledge the support from the National Natural Science Foundation of China (No. 51778557) and the Qing Lan Project (No. 20160512) by the Jiangsu Province Government.

References

Adam D., Markiewicz R.: Energy from earth-coupled structures, foundations, tunnels and sewers. Géotechnique **59**(3), 229–236 (2009). https://doi.org/10.1680/geot.2009.59.3.229

Bardet, J.P.: Bounding surface plasticity model for sands. J. Eng. Mech. **112**(11), 1198–1217 (1986). https://doi.org/10.1061/(ASCE)0733-9399(1986)112:11(1198)

Bourne-Webb, P.J., Amatya, B., Soga, K., Amis, T., Davidson, C., Payne, P.: Energy pile test at Lambeth College, London: geotechnical and thermodynamic aspects of pile response to heat cycles. Géotechnique **59**(3), 237–248 (2009). https://doi.org/10.1680/geot.2009.59.3.237

Brandl, H.: Energy foundations and other thermo-active ground structures. Geotechnique **56**(2), 81–122 (2006). https://doi.org/10.1680/geot.2006.56.2.81

Di Donna, A., Laloui, L.: Numerical analysis of the geotechnical behaviour of energy piles. Int. J. Numer. Anal. Meth. Geomech. **39**(8), 861–888 (2015). https://doi.org/10.1002/nag.2341

Goode III J.C., Zhang M., McCartney J.S.: Centrifuge modelling of energy foundations in sand. In: ICPMG2014—Physical Modelling in Geotechnics: Proceedings of the 8th International Conference on Physical Modelling in Geotechnics 2014 (ICPMG2014), Perth, Australia, 14–17 January 2014, pp. 729–735 (2014). https://doi.org/10.1201/b16200-100

Kalantidou, A., Tang, A.M., Pereira, J.M., Hassen, G.: Preliminary study on the mechanical behaviour of heat exchanger pile in physical model. Géotechnique **62**(11), 1047–1051 (2012). https://doi.org/10.1680/geot.11.T.013

Laloui, L., Nuth, M., Vulliet, L.: Experimental and numerical investigations of the behaviour of a heat exchanger pile. Int. J. Numer. Anal. Meth. Geomech. **30**(8), 763–781 (2006). https://doi.org/10.1002/nag.499

McCartney J.S., Rosenberg J.E. (2011) Impact of heat exchange on side shear in thermo-active foundations. In: Geo-Frontiers 2011: advances in geotechnical engineering pp. 488–498. https://doi.org/10.1061/41165(397)51

Ng, C.W.W., Shi, C., Gunawan, A., Laloui, L.: Centrifuge modelling of energy piles subjected to heating and cooling cycles in clay. Geotechnique Lett. **4**(4), 310–315 (2014a). https://doi.org/10.1680/geolett.14.00063

Ng, C.W.W., Shi, C., Gunawan, A., Laloui, L., Liu, H.L.: Centrifuge modelling of heating effects on energy pile performance in saturated sand. Can. Geotech. J. **52**(8), 1045–1057 (2014b). https://doi.org/10.1139/cgj-2014-0301

Olgun, C.G., Ozudogru, T.Y., Arson, C.F.: Thermo-mechanical radial expansion of heat exchanger piles and possible effects on contact pressures at pile-soil interface. Geotechnique Lett. **4**, 170–178 (2014). https://doi.org/10.1680/geolett.14.00018

Wang, B., Bouazza, A., Rao, M.S., Haberfield, C., Barry-Macaulay, D., Baycan, S.: Posttemperature effects on shaft capacity of a full-scale geothermal energy pile. J. Geotechnical, Geoenvironmental Eng. **141**(4), 04014125 (2015). https://doi.org/10.1061/(ASCE)GT.1943-5606.0001266

Yavari, N., Tang, A.M., Pereira, J.M., Hassen, G.: Experimental study on the mechanical behaviour of a heat exchanger pile using physical modelling. Acta Geotech. **9**(3), 385–398 (2014). https://doi.org/10.1007/s11440-014-0310-7

Effect of Material Stiffness Variation on Shakedown Solutions of Soils Under Moving Loads

Shu Liu[1], Juan Wang[1(✉)], Dariusz Wanatowski[2], and Hai-Sui Yu[2]

[1] Ningbo Nottingham New Materials Institute, University of Nottingham Ningbo China, Ningbo, China
juan.wang@nottingham.edu.cn

[2] Faculty of Engineering, School of Civil Engineering, University of Leeds, Leeds, UK

Abstract. Shakedown limits of pavements and railway foundations can be calculated based on shakedown theorems. These values can be used to guide the thickness designs of pavement and railway constructions considering material plastic properties. However, most existing shakedown analyses were carried out by assuming a unique stiffness value for each material. This paper mainly concentrates on the influence of stiffness variation on the shakedown limits of pavements and railway foundations under moving loads. Finite element models as well as a user-defined material subroutine UMAT are first developed to obtain the elastic responses of soils considering a linearly increasing stiffness modulus with depth. Then, based on the lower-bound shakedown theorem, shakedown solutions are obtained by searching for the most critical self-equilibrated residual stress field. It is found that for a single-layered structure, the rise of a stiffness changing ratio will give a larger shakedown limit; and the increase is more pronounced when the friction angle is relatively high. For multi-layered pavement and railway systems, neglecting the stiffness variation may overestimate the capacity of the structures.

1 Introduction

Most existing pavement and railway design approaches were developed based on empirical data or elastic theory. However, contribution of material plastic properties to the capacity of pavements or railway foundations was not taken into account. Also, the long-term behavior of pavements and railway foundations subjected to repeated moving traffic loads was not well considered in most of the existing design approaches. Many field and laboratory results have demonstrated that when a soil structure is subjected to a moving or cyclic load whose magnitude is larger than the yield limit but smaller than another limit, no further deformation can be observed after some permanent deformation in the first number of load cycles (e.g. Larew and Leonards 1962; Lekarp and Dawson 1998; Werkmeister 2004; Brown et al. 2012 etc.). This phenomenon is known as shakedown and the limit load is termed as shakedown limit. By

W. Steyn et al. (eds.), *Pavement Materials and Associated Geotechnical Aspects of Civil Infrastructures*, Sustainable Civil Infrastructures,
https://doi.org/10.1007/978-3-319-95759-3_6

introducing the shakedown concept, long-term responses of pavements and railway foundations to moving traffic load can be predicted. Shakedown limit, therefore, can be considered as a more rational design load for pavement and railway design against excessive settlement.

Shakedown limits can be determined directly using Melan's (1938) static shakedown theorem (e.g. Sharp and Booker 1984; Raad et al. 1988; Yu and Hossain 1998; Krabbenhøft 2007; Yu and Wang 2012; Wang and Yu 2013a, 2014) or Koiter's kinematic shakedown theorem (e.g. Collins and Clifffe 1987; Boulbibane and Weichert 1997; Li and Yu 2006). As Melan's static shakedown theorem satisfies internal equilibrium equations and stress boundary conditions, it provides a lower bound to the true shakedown limit; therefore it is also named as lower bound shakedown theorem. Koiter's kinematic shakedown theorem satisfies compatibility condition for plastic strain rate and boundary conditions for velocity and therefore it can be used to predict the upper bound of the true shakedown limit. An advantage of the shakedown approach based on these two fundamental shakedown theorems is that the details of the successive elastic-plastic stress fields are not required. Besides, some shakedown solutions have been verified by using numerical step-by-step analyses (e.g. Wang and Yu 2013b; Liu et al. 2016).

Pavement and railway systems are layered structures with diverse material properties in those layers. Even within a single type of material, the material property may also vary at different locations. Typically, the stiffness of soil increases with depth. Stiffness variation with depth has been considered for solving footing problems (e.g. Boswell and Scott 1975; Stark and Booker 1997). However, most of the existing shakedown solutions for pavements and railway foundations were conducted by assuming homogenous materials. In the present study, by assuming a quasi-static response of pavement and railway systems to traffic loads, the effect of material stiffness variation with depth on the shakedown limits will be assessed by using a lower bound shakedown approach.

2 Problem Definition

This study considers that a pressure repeatedly moves on the surface of a three-dimensional half-space along one direction (x-direction). The half-space is made of layers of elastic-perfectly plastic materials. The elastic behavior is described by Young's modulus E and Poisson's ratio v, while the Young's modulus varies linearly with depth z for one material:

$$E_n(\mathrm{z}) = E_{0n}(1 + \mathrm{z}\beta_n) \tag{1}$$

where E_{0n} indicates the Young's modulus on the top of nth layer; $E_n(\mathrm{z})$ indicates the Young's modulus at depth z of the nth layer and β_n is defined as stiffness variation ratio. The plastic behaviors of the materials obey the Mohr-Coulomb model.

The present study will investigate the shakedown limit for two cases considering the effect of stiffness variation. The first case assumes a contact pressure moves on a pavement. The second one considers a train load-induced pressure travels on a railway foundation.

3 Lower Bound Shakedown Analysis

Melan's static shakedown theorem states that an elastic-perfectly plastic structure under cyclic or variable loads will shakedown if a self-equilibrated residual stress field exists such that its superposition with a load induced elastic stress field does not exceed the yield criterion anywhere in the structure. It means three components are essential for the calculation of the shakedown limit, which are elastic stress field, residual stress field and yield criterion. In this study, the Mohr-Coulomb criterion is assumed for all materials. For the elastic stress field in the three-dimensional layered system, numerical calculation is required to obtain solutions. In this study, finite element analyses using the commercial software ABAQUS are conducted. The stress-strain relation (Eq. 2) considering the change of Young's Modulus with depth are programmed into a UMAT subroutine and integrated with the software.

$$\begin{bmatrix}\delta\varepsilon_{xx}\\ \delta\varepsilon_{yy}\\ \delta\varepsilon_{zz}\\ \delta\varepsilon_{xy}\\ \delta\varepsilon_{xz}\\ \delta\varepsilon_{yz}\end{bmatrix}=\begin{bmatrix}1/(E_0+E_0\beta z) & -\upsilon/(E_0+E_0\beta z) & -\upsilon/(E_0+E_0\beta z) & & & \\ -\upsilon/(E_0+E_0\beta z) & 1/(E_0+E_0\beta z) & -\upsilon/(E_0+E_0\beta z) & & & \\ -\upsilon/(E_0+E_0\beta z) & -\upsilon/(E_0+E_0\beta z) & 1/(E_0+E_0\beta z) & & & \\ & & & (1+\nu)/(E_0+E_0\beta z) & & \\ & & & & (1+\nu)/(E_0+E_0\beta z) & \\ & & & & & (1+\nu)/(E_0+E_0\beta z)\end{bmatrix}\begin{bmatrix}\delta\sigma_{xx}\\ \delta\sigma_{yy}\\ \delta\sigma_{zz}\\ \delta\sigma_{xy}\\ \delta\sigma_{xz}\\ \delta\sigma_{yz}\end{bmatrix} \quad (2)$$

Two models are established for the pavement case and the railway case respectively. In both cases, only half of the models are simulated in which x-axis represents the travel direction and the plane y = 0 is the symmetric plane. All elements are chosen as C3D20R which stands for Continuum, three-dimensional, 20-noded, reduced-integrated element. The pavement model consists of an asphalt layer, a subbase layer and subsoil, as shown in Fig. 1, where h_n respects the thickness of the *n*th layer. The wheel-pavement contact load (*P*) distributes within a circle of radius *a* with a maximum pressure $p_{\max} = 3P/2\pi a^2$, which can be formulated as:

$$p = \frac{p_{\max}}{a}\left(a^2 - x^2 - y^2\right)^{1/2} \quad (3)$$

The railway model is shown in Fig. 2 considering the supporting structure for a typical Rheda 2000 single track. Four axle loads belonging to two adjacent bogies on two carriages are considered. The equivalent reaction modulus (k_{sub}) for this system is estimated based on Vesic (1961)'s method and therefore the vertical stress distribution on the top of this structure can be determined based on the beam on elastic foundation theory using the determined k_{sub} value. In the transverse direction (y), the pressure is assumed to be uniformly distributed over y = 0 to y = 1.7 m, because a concrete base with a half-width of 1.7 m is considered to be located on the top. The pressure

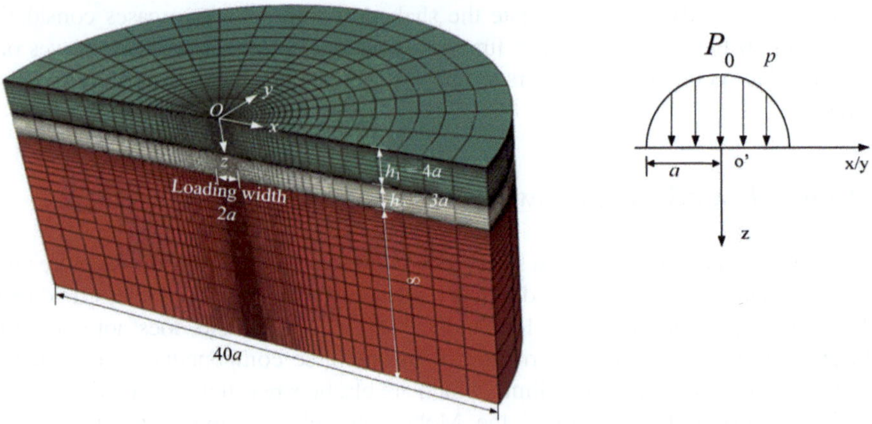

Fig. 1. Load distribution and finite element model for a pavement

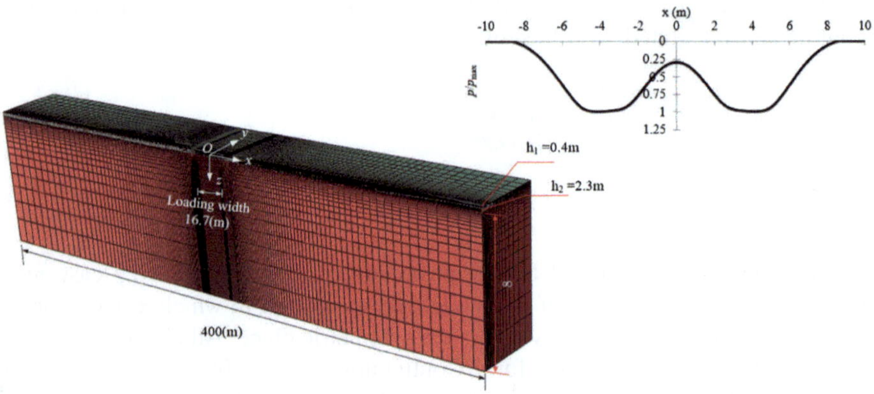

Fig. 2. Load distribution and finite element model for a railway foundation

distribution used in this study is shown in Fig. 2 where $p_{\max}$ indicates the maximum pressure. The load distributions are applied on the models using another user subroutine DLOAD.

In order to obtain the residual stress field, the critical residual stress fields of Yu and Wang (2012) will be used, as formulated in Eq. 4. These critical residual stresses are obtained by making sure the total stress state of one point just touches the Mohr-Coulomb yield surface at each depth z = j while fulfilling the self-equilibrium and boundary conditions.

$$
\begin{aligned}
\sigma^{r}_{xx-\min} &= \min_{z=j}\left(-M_i + \sqrt{-N_i}\right) \\
\sigma^{r}_{xx-\max} &= \max_{z=j}\left(-M_i - \sqrt{-N_i}\right)
\end{aligned}
\qquad (4)
$$

with

$$\mathrm{M} = \lambda\sigma^{e}_{xx} - \lambda\sigma^{e}_{zz} + 2\tan\phi_{n}(c_{n} - \lambda\sigma^{e}_{zz}\tan\varphi_{n}),$$
$$\mathrm{N} = 4(1 + \tan^{2}\phi_{n})[(\lambda\sigma^{e}_{xz})^{2} - (c_{n} - \lambda\sigma^{e}_{zz}\tan\phi_{n})^{2}].$$

in which i represents a general point at depth z = j; λ is a dimensionless load factor; ϕ is material friction angle, c is material cohesion; σ^{e}_{ij} is elastic stress field induced by a unit p_{max}; the subscript n (n = 1, 2, 3…) means the nth layer. Tension positive notation is applied throughout this paper.

By substituting the elastic stress fields and either of the critical residual stress fields into the Mohr-Coulomb yield criterion $f(\sigma) \leq 0$, the present shakedown problem is presented as a mathematical optimisation problem:

$$\max \lambda$$
$$\text{s.t.}\begin{cases} f(\sigma^{r}_{xx}(\lambda\sigma^{e}), \lambda\sigma^{e}) \leq 0, \\ \sigma^{r}_{xx}(\lambda\sigma^{e}) = \sigma^{r}_{xx-\max} \text{ or } \sigma^{r}_{xx-\min}. \end{cases} \tag{5}$$

The mathematical optimization process is programmed using MATLAB. For each layer, one maximum admissible λ could be found, marked as λ^{n}_{sd} and therefore $\lambda^{n}_{sd}p_{\max}$ is the shakedown limit of the nth layer. The minimum value among all them $\lambda^{n}_{sd}p_{\max}$ is then recorded as the shakedown limit of the whole structure.

4 Results and Discussions

4.1 Model Verification

A single-layered problem is first investigated by giving identical material properties (E_0 = 100 MPa, v = 0.3, β = 0) to all layers. The results are compared with the analytical solutions of Wang and Yu (2013a) and show good agreements (Fig. 3).

Figure 4 also demonstrates the effect of the stiffness variation ratio on the shakedown limit of the single-layered structure. It indicates that the shakedown limit increases with the stiffness variation ratio and the change is more pronounced when the friction angle is high.

4.2 Pavement Case

Typical material properties are shown in Table 1 for a pavement structure considering a temperature of 35 °C. If all the materials are with constant stiffness, the shakedown limit of the layered structure is 296 kPa and failure will initiate in the second layer. Figure 5 demonstrates the effect of the stiffness variation ratio (β) in one layer on the shakedown limits when the stiffness variation ratio in the other two layers are assumed to be equal to 1. For all the cases considered here, the second layer is always the most critical layer and shakedown limits of the structure could be either smaller or larger than 296 kPa. Moreover, the increase of β_1 and β_3 leads to a larger shakedown limit of

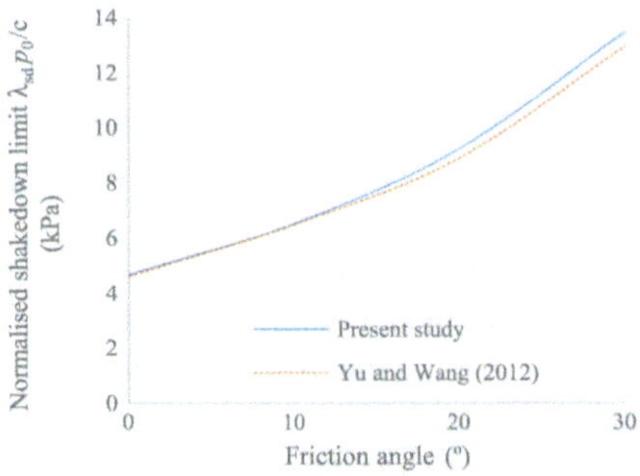

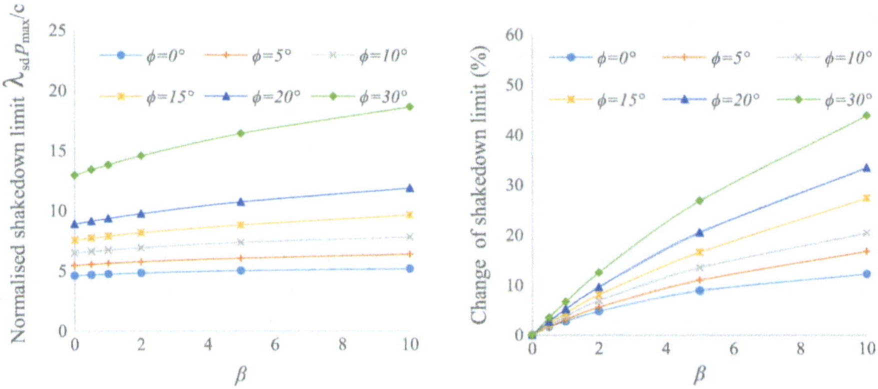

Fig. 3. Effect of stiffness variation ratio on the shakedown limits of a single-layered soil structure

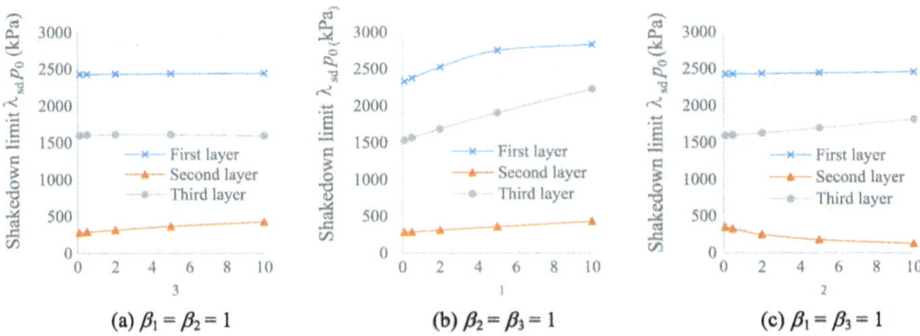

Fig. 4. Effect of stiffness variation ratio on shakedown limits of a pavement structure

Table 1. Material properties for a pavement

	E_{0n} (MPa)	ν_n	h_n (m)	c_n (kPa)	ϕ_n (°)
Asphaltic layer	690	0.3	0.4	300	30
Subbase	150	0.3	0.3	2	40
Subsoil	60	0.4	∞	10	0

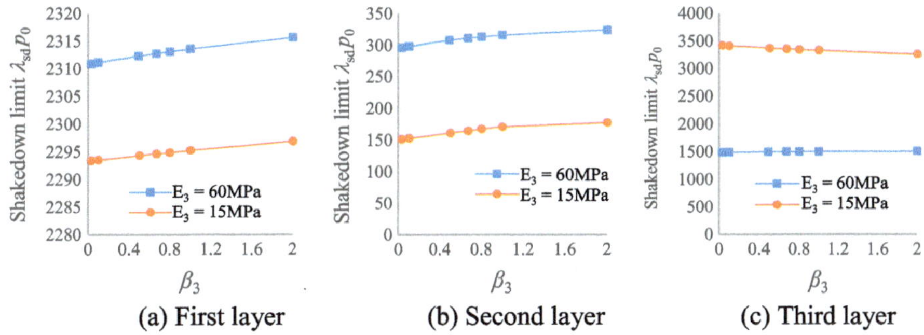

Fig. 5. Effect of β_3 on shakedown limits of pavement structures when $\beta_1 = \beta_2 = 0$

the second layer, while a reduction in the shakedown limit can be observed along with the rise of β_2. Besides, another two cases considering E_3 = 15 MPa and 60 MPa are studied by assuming the stiffness moduli in the first two layers are constant with depth, while the stiffness variation ratio in the third layer varies from 0.03 to 2 (Hammam and Eliwa 2013; Rowe and Booker 1980, 1981). It can be known that failure happens in the second layer first. For the case with stiff subsoil, shakedown limits of all those three layers rise with the increase of β_3. For the case with soft subsoil, the shakedown limit of the third layer decreases with the rising β_3, whereas those of the other two layers are both increased. When a typical stiffness variation ratio 0.67 (Rowe and Booker, 1980) is taken for the subsoil, the pavement shakedown limits are increased by 5.6% for the stiff case and 22.5% for the soft case respectively comparing with the situations with constant stiffness assumption.

4.3 Railway Case

Concerning a typical three-layered track system whose material properties are given in Table 2, the second layer will fail prior to the other two layers for the constant stiffness case. The shakedown limit of the layered structure is 22.43 kPa. Figure 6 illustrates the effect of the stiffness variation ratio on the shakedown limits. It can be seen that the second layer is still the weakest layer. In addition, it is found that the increasing stiffness variation ratio in one layer will result in a smaller shakedown limit in that layer but larger shakedown limits in the other two layers. Comparing with the constant stiffness case, changes of the stiffness variation ratios can lead to either larger or smaller shakedown limit. For instant, larger shakedown limits can be observed when β_3 is larger than 3 (Fig. 6a) or β_2 is smaller than 0.5 (Fig. 6c). Another case is considered by

Table 2. Material properties for a railway foundation

	E_{0n} (MPa)	v_n	h_n (m)	c_n (kPa)	ϕ_n (°)
Anti-frozen layer	200	0.3	0.4	1	45
Subgrade bed	150	0.3	2.3	2	35
Subsoil	60	0.4	∞	10	0

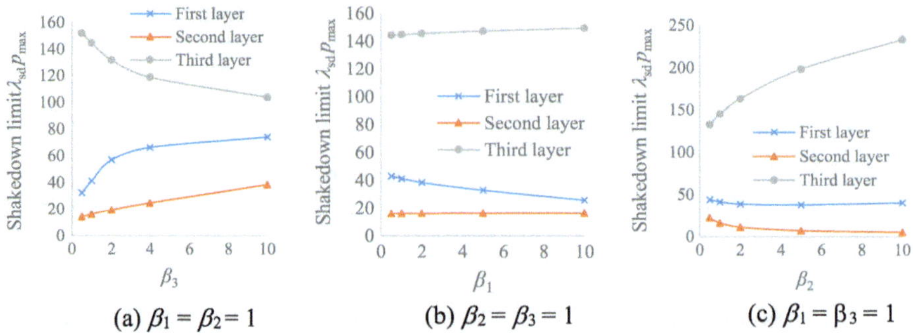

Fig. 6. Effect of stiffness variation ratio on shakedown limits of a railway foundation when c_3 = 10 kPa

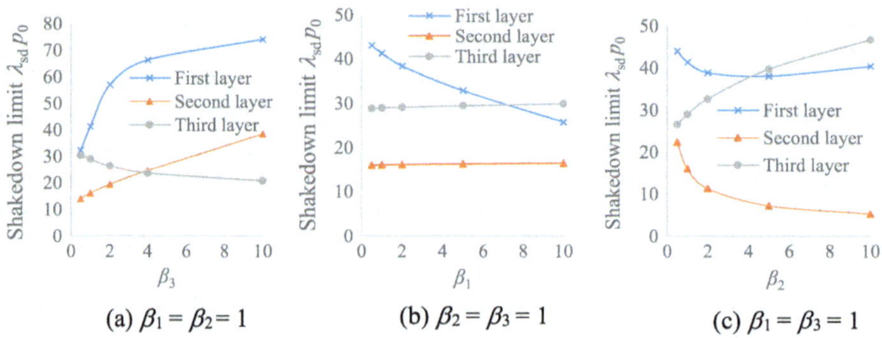

Fig. 7. Effect of stiffness variation ratio on shakedown limits of a railway foundation when c_3 = 2 kPa

decreasing the cohesion of the third layer from 10 kPa to 2 kPa. As shown in Fig. 7, failure could initiate in the second layer or the third layer. For the case considered in Fig. 7a, the shakedown limit is maximized when $\beta_3 = 3.7$. If $\beta_3 < 3.7$, the second layer is critical; otherwise failure occurs in the third layer.

5 Conclusions

(1) The lower bound shakedown approach has been adopted to solve pavement and railway problems considering a linearly increasing stiffness modulus with depth. Finite element models and a user subroutine UMAT have been developed to obtain the elastic responses of soils. The results are well validated.
(2) For the single-layered problems, the rise of the stiffness variation ratio leads to a larger shakedown limit. More obvious changes can be observed when the friction angle is high.
(3) For the typical layered pavement system, the second layer is the most critical layer. The shakedown limit of the layered system increases with the rise of β_1 and

β_3 as well as the decrease of β_2. Moreover, the subsoil stiffness has an obvious influence on the trend of the shakedown limit with the stiffness variation ratio.

(4) For the typical railway system, the second layer is also the most critical layer. Increasing stiffness variation ratio in one layer results in a smaller shakedown limit in that layer, but larger shakedown limits in the other two layers. Comparing with the constant stiffness cases, the consideration of material stiffness variation lead to either larger or smaller shakedown limit of the railway foundation. Considering a case with a smaller subsoil cohesion, the failure mode may change from the second layer failure to the third layer failure.

References

Boulbibane, M., Weichert, D.: Application of shakedown theory to soils with non-associated flow rules. Mech. Res. Commun. **24**(5), 513–519 (1997)

Brown, S.F., Yu, H.S., Juspi, H., Wang, J.: Validation experiments for lower-bound shakedown theory applied to layered pavement systems. Géotechnique **62**(10), 923–932 (2012)

Collins, I.F., Cliffe, P.F.: Shakedown in frictional materials under moving surface loads. Int. J. Numer. Anal. Meth. Geomech. **11**(4), 409–420 (1987)

Hammam, A.H., Eliwa, M.: Comparison between results of dynamic and static moduli of soil determined by different methods. HBRC J. **9**(2), 144–149 (2013)

Krabbenhøft, K., et al.: Shakedown of a cohesive-frictional half-space subjected to rolling and sliding contact. Int. J. Solids Struct. **44**(11–12), 3998–4008 (2007)

Larew, H.G., Leonards, G.A.: A strength criterion for repeated loads. Highway Res. Board Proc. **41**, 529–556 (1962)

Lekarp, F., Dawson, A.: Modelling permanent deformation behaviour of unbound granular materials. Constr. Build. Mater. **12**(1), 9–18 (1998)

Li, H.X., Yu, H.S.: A nonlinear programming approach to kinematic shakedown analysis of frictional materials. Int. J. Solids Struct. **43**(21), 6594–6614 (2006)

Liu, S., et al.: Shakedown solutions for pavements with materials following associated and non-associated plastic flow rules. Comput. Geotech. **78**, 218–266 (2016)

Raad, L., et al.: Stability of multilayer systems under repeated loads. Transp. Res. Rec. **1207**, 181–186 (1988)

Rowe, R.K., Booker, J.R.: The behaviour of footings resting on a non-homogeneous soil mass with a crust. Part II. Circular footings. Can. Geotech. J. **18**(18), 265–279 (1980)

Rowe, R.K., Booker, J.R.: The elastic displacements of single and multiple underream anchors in a Gibson soil. Geotechnique **31**(1), 125–142 (1981)

Sharp, R.W., Booker, J.R.: Shakedown of pavements under moving surface loads. J. Transp. Eng. ASCE **110**, 1–14 (1984)

Vesic, A.B.: Bending of beams resting on isotropic elastic solid. J. Eng. Mech. **87**(2), 35–54 (1961)

Wang, J., Yu, H.S.: Shakedown analysis and design of layered road pavements under three-dimensional moving surface loads. Road Mater. Pavem. Des. **14**, 703–722 (2013a)

Wang, J., Yu, H.S.: Residual stresses and shakedown in cohesive-Frictional half-space under moving surface loads. Geomech. Geoeng. **8**(1), 1–14 (2013b)

Wang, J., Yu, H.S.: Three-dimensional shakedown solutions for anisotropic cohesive-frictional materials under moving surface loads. Int. J. Numer. Anal. Meth. Geomech. **38**(4), 331–348 (2014)

Werkmeister, S., et al.: Pavement design model for unbound granular materials. Jo. Transp. Eng. **130**(5), 665–674 (2004)

Yu, H.S., Hossain, M.Z.: Lower bound shakedown analysis of layered pavements using discontinuous stress fields. Comput. Methods Appl. Mech. Eng. **167**(3–4), 209–222 (1998)

Yu, H.S., Wang, J.: Three-dimensional shakedown solutions for cohesive-frictional materials under moving surface loads. Int. J. Solids Struct. **49**(26), 3797–3807 (2012)

Social Considerations in Selection of Sustainable Pavement Designs

Sundeep Inti[1], Megha Sharma[2(✉)], and Vivek Tandon[2]

[1] Mixonsite USA Inc, Buffalo Grove, IL, USA
sundeep@mixonsite.com
[2] The University of Texas at El Paso, Texas, USA
msharma4@miners.utep.edu, vivek@utep.edu

Abstract. To assess the sustainability of pavement, one needs to estimate the economy, environment, and social consequences during its service life. Although Life Cycle Assessment and Life Cycle Cost Analysis tools are available for evaluating the environmental and economic impact of pavement sustainability, little or no attention has been focused on evaluating the social component of pavement sustainability. It can be attributed to lack of an established framework and unavailability of data needed for analysis. In this study, a framework for performing Social Life Cycle Assessment (SLCA) is proposed. To evaluate a social component of sustainability, the traffic noise, traffic delay costs, and vehicle operating costs are considered as social indicators. These indicators help in estimating the impact of pavement on surrounding neighborhood and road users. In the end, a case study was performed to estimate the influence of social impacts on sustainability. Based on the results, it can be concluded that inclusion of social indicators is crucial and should be included in the sustainability assessment.

Keywords: Sustainability · Social life cycle assessment · Social impact indicators

1 Introduction

Sustainability is an emerging science, and constant efforts are being made to incorporate sustainability principles into highway construction. Federal Highway Administration (FHWA) (2015), defines "a sustainable highway should satisfy functional life cycle requirements of societal development and economic growth while striving to enhance the natural environment and reduce consumption of natural resources." Developing a sustainable highway with no adverse consequences (on Mother Nature and society) and with full economic benefit is far from reality. Sustainability is a relative concept and engineers need to seek to minimize the impacts in all possible ways. Construction of highways requires an enormous amount of natural resources, capital, workforce, etc., and pavement design drives these parameters. Design and planning engineers have an opportunity to develop a sustainable design by systematically evaluating all the possible alternatives and select the one which most balances the sustainable factors.

W. Steyn et al. (eds.), *Pavement Materials and Associated Geotechnical Aspects of Civil Infrastructures*, Sustainable Civil Infrastructures,
https://doi.org/10.1007/978-3-319-95759-3_7

Many researchers (Anderson and Muench 2013; Spangenberg 2013; Lew et al. 2016) stated that incorporation of sustainability principles at the beginning of the project like in the design and planning phases would contribute significantly towards sustainable development. A systematic and robust assessment is needed to measure sustainability. Life cycle assessments are widely used tools for measuring sustainability because they evaluate the pavements from the initial construction through the disposal. They are especially useful for comparing designs or products. Economic and environmental impacts of pavement design have been traditionally evaluated through Life Cycle Cost Analysis (LCCA) and Life Cycle Assessments (LCA) tools, respectively. However, the evaluation of social impacts in life cycle approach is still nascent. The authors are not stating that the highway projects are not considering social implications at all. But arguing that there is an enormous scope of decisions with material or design selection that can be helpful in reducing impacts on society. Henceforth, the main focus of this study is to showcase assessment of various social implications that are related to pavement designs. The present study mainly emphasizes on highways in urban areas where the significant population is affected due to pavements. Since life cycle assessments considered as one of the widely accepted evaluation tools, the following section examines the current status of social life cycle assessment.

2 Social Life Cycle Assessment (SLCA)

Social Life Cycle Assessment (SLCA) is a systematic process which uses the best available science to collect best available data to report on social impacts (positive and adverse) in product life cycles from extraction (cradle) to final disposal (grave) (Benoît et al. 2010). Most of the published literature on social sustainability has been focused on products while paying minimal attention to services. Social Hotspots Database (SHDB) is a resource for performing SLCA (Benoit et al. 2013). More than a hundred products are evaluated using SHDB (like a laptop, Orange Juice, etc.), and nearly two hundred organizations have been using SHDB for performing SLCA for decision-making. However, usage of SHDB is very minimal in construction. The following are some of the challenges are faced in conducting SLCA in road design selection:

- According to Dubois and Gadde (2002), construction is a loosely coupled system (less control over the supply chain) whereas mass production manufacturing is a tightly coupled system (tight control over the supply chain). The construction industry is complicated, and the sources of complexity are numerous due to uncertainty and interdependence among tasks (Gidado 1996). Construction is not a single industry but an "industry of industries" which requires additional foundational work apart from traditional industrial approaches (Fenandez Solis). For example, typically the use phase in SLCA has been ignored (in the Manufacturing sector) as most researchers agree that the use phase is hard to evaluate. However, in highways, the impacts on local communities will be more during use phase than any other phases and neglect it is inappropriate.

- Accounting all social issues like creation of local jobs, impacts on the health and education of local communities, disruption of community social behaviors, workforce safety, and well-being, human rights, etc., is impractical (Schmidt et al. 2004) and often these impacts are not explicitly related to pavement designs.
- SLCA requires multi-discipline and multi-dimensional approach suggesting the involvement of professionals from various disciplines like psychology, environmental sociology, engineering, etc., to bridge the gaps in theoretical concepts about social impacts.
- There is a lack of standard procedure and guidelines for conducting SLCA of pavements due to non-availability of databases.
- Sometimes it is a challenging task to identify and quantify social impacts due to high capital and time requirement.
- Lack of clear social impact indicators (social impact indicator acts as a measure of social implications) about pavement design choice.

3 Research Motivation and Objectives

There is an opportunity for design engineers and transportation managers to minimize the social impacts through the pavement design selection. The following are the objectives of this study:

- To develop a framework and identify social impact indicators for performing SLCA of alternative pavement structure designs.
- To demonstrate the proposed framework and employ the defined social impact indicators through a case study.

4 Objective 1: Framework for Performing SLCA for Pavement Structure Designs

The essence of SLCA is to compute effects on people due to a pavement throughout its life. An approach was proposed to perform SLCA for highways upon reviewing a thorough literature in other industries. Various researchers (Benoît et al. 2010; Benoît 2010; Hunkeler 2006; Paragahawewa et al. 2009; Weidema 2005) have suggested using an approach like LCA. However, SLCA differs from LCA because SLCA demands to set more priority on the stakeholders (Grießhammer et al. 2006). Project stakeholders are the people who have a concern for the success of the project and the environment in which the project operates (Olander 2006). Figure 1 shows the proposed framework for SLCA. The key phases (Goal and Scope, Inventory Analysis, Impact Assessment, and Interpretation) are similar to LCA. The fundamental difference is the assessment focusses estimating effects on people.

The impacts on stakeholders due to a highway in its lifetime are numerous and assessing all of them is not possible. To make the SLCA a feasible study, one should explicitly define the key stakeholders and scope of the assessment. For example, the

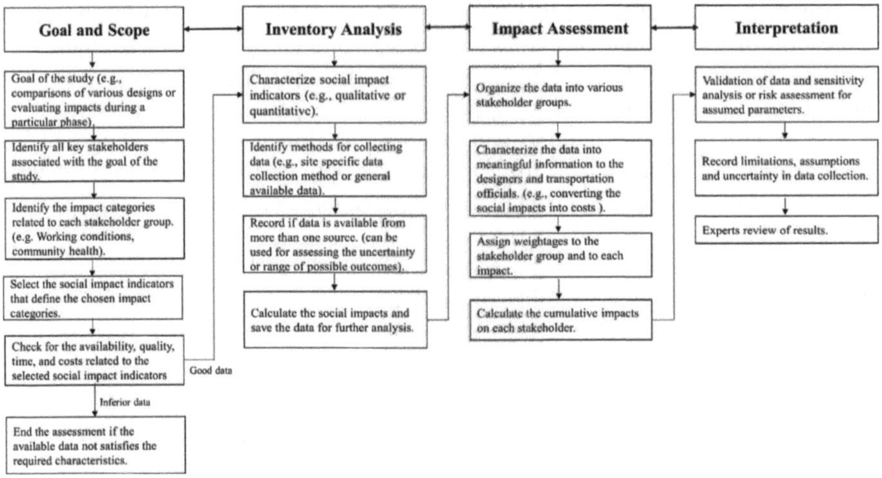

Fig. 1. Proposed SLCA framework

goal for a highway agency is to identify the impacts on the local community. Also, one should consider the time, cost, quality, and availability of data about social implications on the selected stakeholders. The impacts on the stakeholders need to be assessed/collected during the inventory analysis. For example, if SLCA is conducted to estimate the effects on users, it may determine the costs incurred (like traffic delays, fuel consumption) to users due to each pavement structure design. The purpose of impact assessment is to convert the inventory data into meaningful information. In the interpretation phase, the limitations, assumptions, risks, etc., are provided to designers so that they can assimilate the background of the study.

5 Objective 2: Social Impact Indicators Related to Pavement Structure Design Selection

Impact categories describe the effect on people due to pavement structure design. For example, if assessing the workforce then safety and health, labor rights, etc., are some of the impact categories. The impact categories can have subcategories. For instance, labor rights may have subcategories like child labor, gender equity, forced labor, etc. Social impact indicators are used to measure impact categories (or subcategories). The number of child labor incidents observed in an organization (or geographical location) is a possible social impact indicator for child labor. According to Weidema (2006), the requirement for a good social indicator is that it permits quantification of the extent (incidence or prevalence) as well as duration and severity of the considered impact. Social indicators should possess a valid relation with the impact category. Usually, pavement structures are designed for 30–50 years of life in which the construction and maintenance activities will be around 3–5 years. The local community and highway users are involved in most pavements life. The focus of the study is to select to design which has fewer impacts on the local community and road users.

Impacts on Local Communities

Upon appraising the published literature on social sustainability, SLCA, and other tools to compute social consequences "traffic noise" is considered as a social impact indicator in this study. Traffic noise can be measurable and is related to negative impacts on local communities.

Why can Traffic Noise be a Social Impact Indicator?

Noise relates to the exposure of a target to the unwanted sound produced by a source Cucurachi et al. (2014). Noise-induced effects on humans are numerous such as acoustic trauma, reduced efficiency, difficulty in concentrating, hearing loss, increased blood pressure, hypertension, etc., (OSHA 2015). It may also cause speech interference, sleep disturbance, annoyance and a loss of the quality of life (Bernhard 2005). In addition to the health impacts, noise affects real estate and property values. It is apparent that sound affects the human health and well-being of the community.

How can Traffic Noise from Pavements be a Social Indicator?

Noise emissions from pavements during all phases except use phase are from machinery and equipment. During use phase, the sources of highway traffic noise are from vehicles engine/drivetrain, exhaust, aerodynamic, and tire/pavement interaction noise (Bernhard 2005). Tire-pavement interaction noise is predominant than all other sources for cars and trucks traveling at speeds higher than 31 and 50 MPH, respectively (Muzet 2007). In normal conditions, the average vehicles speed will be greater than 50 mph on highways, so the predominant source of traffic noise is tire-pavement interaction. Different pavement designs have separate surface properties and noise emitted will be distinct. This differentiating factor can be valuable for decision makers in selecting pavements.

How can Traffic Noise from Pavements be Quantified?

To predict highway noise levels, FHWA along with the John A Volpe National Transportation Systems Center, Harris Miller & Hanson, Inc., and Foliage Software Systems, Inc. have developed Traffic Noise Model (FHWA TNM®). TNM can estimate traffic noise from various types of pavement surfaces. The TNM2.5 (latest version of the model) estimates the hourly equivalent noise level in decibels (dbA Leq). Multiple inputs needed for the TNM 2.5 are vehicle speed, a distance of receptor point from the center of the road, ground classification (soft vs. hard ground), and counts of different vehicle types (i.e., passenger cars, medium trucks, and heavy trucks), etc. Also, TNM 2.5 has the capacity in modeling barriers at any location from the roadway to reduce the traffic noise on inhabitants. FHWA has recommended noise abatement criteria for different activity areas (schools, hotels, etc.).

How to Relate Calculated Traffic Noise as Impacts on Humans?

It is evident that noise affects people's health but, exact effect relation between noise levels to the type of ailment (people's health) is not well established. Hence, this study proposes to estimate the additional costs required to noise abatement structures to reduce the traffic noise on humans. Noise comprises of three indivisible traits: source, propagation, and receiver. Three ways to reduce the noise impacts on humans are:

(1) Development of quieter pavements and usage of quieter cars like electric vehicles (EVs) and hybrid electric vehicles (HEVs).
(2) Noise abatement structures like noise barrier walls to break the path of sound propagation for minimizing the impact of noise on receivers.
(3) Soundproof windows, soundproof walls, etc. to reduce noise at receivers.

Out of the three available options, this study utilized option two, i.e., reducing the noise by using noise barriers. The other two choices are users personal favorites and hard to predict. So, the traffic noise impacts can be indirectly quantified as the cost of noise barrier walls required. The increase in the NBW costs suggests higher implications on the neighborhood.

Impacts on Road Users

The performance (development of distresses) of a pavement influences some of the road users' expenses. Vehicle operating costs, safety, and accidents, comfort, and convenience, vehicle tailpipe emissions, etc., by some means, are related to highway performance. Since the distresses generated on each pavement structure design are different from other, the impacts on road users' costs will be distinct. The difference in the operating expenses calculated for the vehicles traveling on different pavements will provide needful information for choosing pavement design. We suggest four such user costs which will help in differentiating the pavement designs.

Traffic Delay Costs

In general, traffic delays are anticipated during maintenance activities of pavement, even though unforeseen conditions like accidents and climatic adversities causes delays they are difficult to include in initial estimates. Each pavement designs have different maintenance strategies. Traffic delays impact users in two ways: (1) additional fuel consumption, (2) loss of time. These costs can be accounted using RealCost2.5 application developed by FHWA.

Vehicle Operating Costs (VOC) (Fuel Consumption, Repair and Maintenance, Tire Wear Costs)

The most important factor that influences these three costs is pavement rolling resistance. Pavement roughness, structural responsiveness, and surface macrostructure weigh-in for pavement rolling resistance (Thomas et al. 2015). VOCs are currently estimated through International Roughness Index (IRI) a measure of pavement roughness. Even though other factors like macro-texture measured as mean profile depth (MPD) and structural responsiveness contributes towards rolling resistance, they are not widespread for the following reasons:

- Surface texture (MPD) do not affect the fuel consumption of any vehicle class except for heavy trucks traveling at lower speeds (Chatti and Zaabar 2012) and its impact compared with IRI is much smaller to the extent that it is statistically insignificant (Thomas et al. 2015).

- Even though structural responsiveness affects the rolling resistance, its influence (through various models) has not been comprehensively validated with any experiment till now.

6 Objective 3: Case Study

To achieve objectives of this study, a study is performed by evaluating four similar pavement designs (Fig. 2) that are expected to provide service for 30 years. Equivalent design (Stephanos 2008) implies that each design alternative is expected to perform equally during its design life (provides the same level of service, over the same performance period, and has similar life-cycle costs). Three similar designs were flexible pavement designs while the fourth pavement design was rigid pavement. The flexible pavement design was developed using FPS 21 (TxDOT design program) while the rigid pavement design was developed using the AASHTO Design Guide. The rigid pavement design consists of Continuously Reinforced Concrete Pavement (CRCP) surface layer. The selected pavement designs are traditionally used with in urban traffic conditions within Texas.

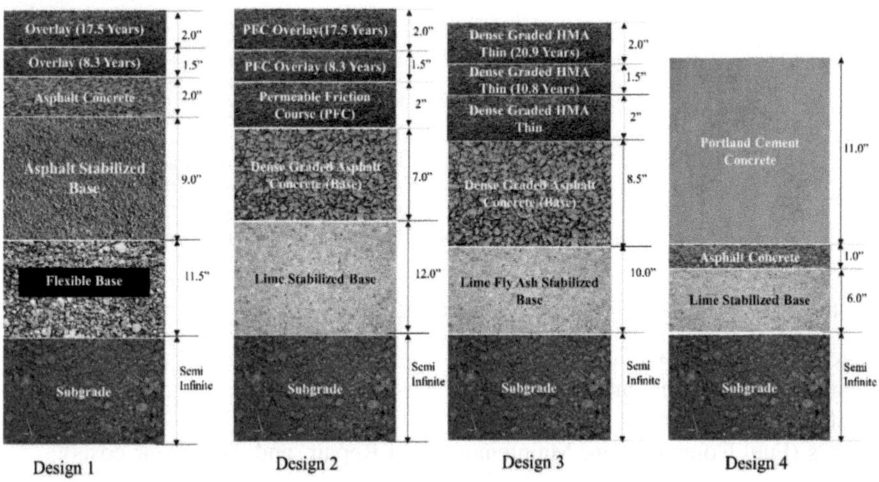

Fig. 2. Selected pavement designs for evaluation

Each of the pavement design is expected to carry Annual Average Daily Traffic (AADT) of 61,236 on each side in 2014 on a six-lane highway (three lanes on each side). The traffic consists of 10.6% of trucks and 89.4% of passenger cars, and annual traffic growth is assumed to be 0.75% for next 30 years. Subgrade conditions were considered the same for all the pavements. The three flexible pavement designs varied in their material composition and thickness of layers. The fourth design is not expected to require maintenance up to 30 years. On average, the flexible pavement design is expected to have two rehabilitations during 30 years of service life. Although rigid pavement is not expected to have major rehabilitation, some minor maintenance (repairing two patches 10' by 20' for every lane-mile at 20 years) for rigid pavement is expected and included in the study. It should be mentioned that rigid as well as flexible pavements may require major rehabilitation because of premature failures, however, were not included in this study. The following sections demonstrate the calculation of impacts on the local community and road users by using the selected social impact indicators and implementing them in the proposed framework. Figure 3 shows the various phases of proposed SLCA and corresponding measures implemented to assess impacts on the local community and road users.

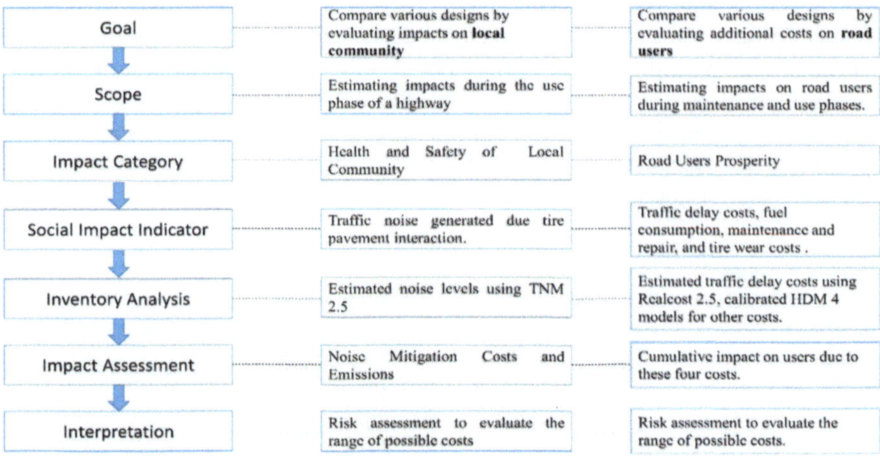

Fig. 3. Various phases of proposed SLCA for assessing pavement designs

In this case study, the following costs are calculated:

a. Agency Construction Costs
b. Traffic Noise Abatement Costs
c. Traffic Delay Costs
d. VOCs (Fuel Consumption, Maintenance and Repair, and Tire Wear costs).

All the construction and maintenance costs in this assessment are from Low Bid Prices of El Paso [Average Low Bid Unit Prices, TxDOT (2016)] and RS Means (2012). Traffic delay values are as per TxDOT 2014 user delay costs ($/Vehicle hour) $21.73 and for a truck is $31.71. The unit prices for fuel consumption, tire costs, and maintenance costs are from NCHRP 720 (2012). Prices are modified based on 2016 consumer price indices. The future costs are considered regarding constant dollars (Inflation not considered) as per the recommendation LCCA Primer FHWA (2002). The probabilistic analysis was used for estimating the net present worth of all costs. Monte Carlo simulations are used to perform the analysis. All the NPWs shown in the following sections are the mean values of 15,000 simulations and are estimates for 30 years for 5 miles of highway. Costs inputs are assumed as normal distribution in most cases. The unit prices discussed above are regarded as mean prices and ten percent of the prices as standard deviations. The discount rate found in the US is usually four percent (*28*). But in the analysis, a triangular distribution with minimum value as three, mean value as four and maximum as seven percent was considered, and the assumptions for the analyses are included in the following sections.

(a) **Agency Construction Costs**

Initially, life cycle costs for each design (pavement construction and maintenance costs) incurred by the agency for 30 years were calculated using FHWA's Realcost2.5. If the design life of the pavement is beyond 30 years, then remaining service life is converted into salvage value and considered as a negative cost. The mean estimated costs [net present worth (NPW)] for five miles are $16.53, $17.70, $15.07, and $17.20 million for Designs one, two, three and four, respectively. At all probabilities, the Design 3 has the lowest construction costs, and Design 2 has the highest.

(b) **Traffic Noise Impacts on Local Community**

The primary assumption of this study is that the highway is surrounded by a residential area (the most common situation in urban areas), and the noise threshold level is 67 dba Leq as per FHWA recommendation. A preliminary assessment was performed to study how traffic noise levels vary with distance say 500 ft from the edge of the highway for all designs for 30 years using TNM2.5. The results indicate that the houses within the vicinity (say 100 ft) of highways were facing higher traffic noise. The noise effects were within the permissible levels at 400, 350, 400, 450 ft away from the road for designs one, two, three, and four, respectively.

Next, an evaluation was conducted for modeling the noise levels by inserting noise barrier walls (NBW) through TNM 2.5. Calculated the height of NBW's required for each pavement design to keep the traffic noise levels within permissible limits at a specified distance from pavement. Table 1 summarizes the height of NBW's needed for various designs to keep the traffic noise below thresholds at different distances from the edge of the pavement. Design 4 generates most traffic noise hence needs higher NBW, and Design 2 requires lowest NBW within the selected designs. Figure 4 shows the predicted noise contours with and without noise barriers using TNM 2.5 for Design 3. It is apparent that the noise levels are above 67 dB up to 400 ft. Construction of NBW of 12 ft reduces the noise levels after 100 ft. A higher height of the wall is required to reduce noise levels further.

In addition to the construction, some periodic maintenance was assumed. The maintenance includes removal of graffiti every year (considered 1% of walls needs graffiti removal at a price of 1$ per square foot), surface maintenance (includes aesthetics of walls expected every ten years at the cost of 1$ per square foot). It also includes replacement of damaged barrier walls (assumed 1% walls needs replacement every five years). The maintenance operations are also included in the probabilistic analysis. The NPW for constructing NBW/mile for various designs at specific distances are displayed in Fig. 5. As anticipated, the NPW is highest for Design 4 and lowest for Design 2.

Table 1. Height of NBW to minimize traffic noise

Design	No barrier	Barrier ht. in ft. required to reduce traffic noise at distance					
	Distance in feet from pavement (noise below 67 db)	50 ft	100 ft	200 ft	300 ft	400 ft	500 ft
1	400	16.0	13.0	9.0	4.0	0.0	0.0
2	350	14.5	12.0	7.0	2.5	0.0	0.0
3	400	16.0	13.0	9.0	4.0	0.0	0.0
4	450	18.0	15.0	10.0	5.0	2.5	0.0

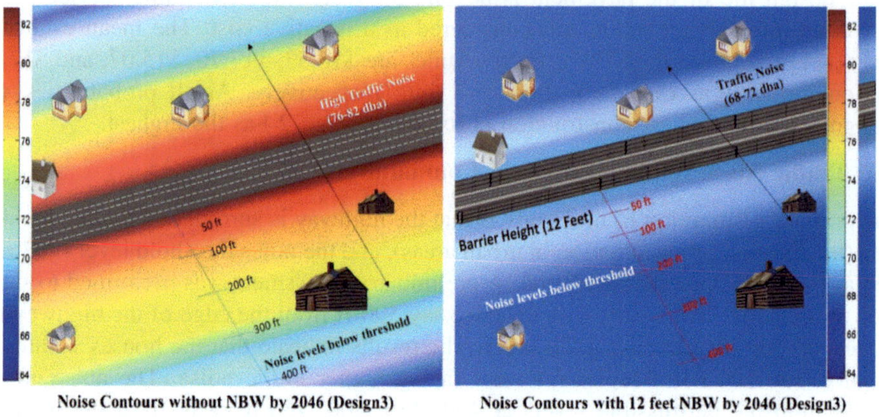

Fig. 4. Predicted traffic noise contours with and without noise barriers for Design 3 by 2046

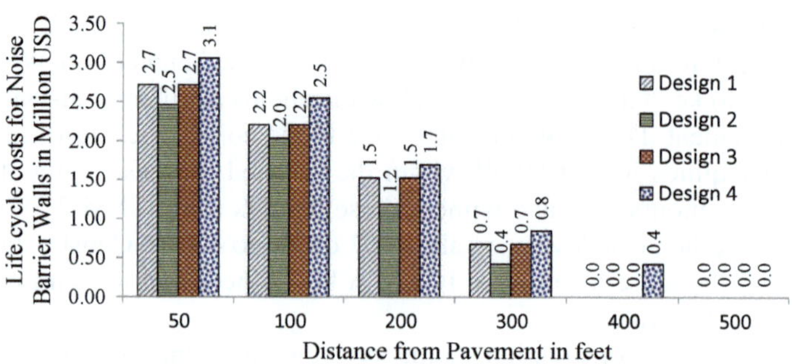

Fig. 5. NPW for a mile of NBW construction

(c) Traffic Delay Costs

The user expenses incurred during maintenance of highways (additional travel time and increased fuel consumption) are estimated using FHWA's Real cost 2.5. The timing of maintenance is the key factor that influences the traffic delay costs. The traffic delay estimates, for various maintenance scenarios (number of maintenance hours and time of maintenance), are summarized in Table 2 for maintenance of five miles highway. We assumed that one lane is closed during maintenance operations. The results indicate maintenance in the daytime and during peak hours of traffic produces more traffic delay costs and maintenance during night times have a very minimal impact on users.

Table 2. Influence of maintenance timing on traffic delay costs

Maintenance timing	Number of maintenance hours per day	Traffic delay costs on user in millions of dollars (Designs)				Remarks
		1	2	3	4	
9AM–5PM	8	4.81	4.84	3.99	0.67	Peak hours of traffic (Daytime)
10PM–8AM	8	0.00	0.00	0.00	0.00	Non-peak hours (Night time)
8AM–8PM	12	8.95	9.00	7.42	1.25	12-h closure of a lane (Day time)
8PM–7AM, 11AM–4PM	16	1.35	1.36	1.12	0.19	Non-peak hours of traffic (Day and night)
24 h	24	6.40	6.44	5.31	0.89	All day maintenance

(d) **Vehicle Operating Costs (Fuel Consumption, Maintenance and Repair, Tire Wear)**

To estimate the VOCs (Fuel Consumption, Maintenance and Repair, and Tire wear) for pavements, NCHRP Report 720 models were employed. These models were calibrated using vehicles (five different vehicle group) driven on roads of the known condition. An additional advantage is that these models reflect the present vehicle technologies in the US. IRI and MPD are the distress that guides these models. The MPD in all models was assumed to be 1 mm and calculated costs based on only IRI data. Some inputs significantly influence these operating cost models like vehicle speed, diesel and gasoline price, the discount rate, tire costs, vehicle group classification within the traffic, etc. The probable range of value of each data point is considered to a probabilistic function. For instance, the speed of the vehicles is considered as a triangular probabilistic distribution with a lower limit, mode, and peak value as 35, 55, and 70 MPH respectively. Typically, the pavements were designed based on truck traffic, and passenger cars do not contribute significantly towards pavement structural damage. However, there is considerable impact on the VOCs of a passenger car due to pavement performance. In urban areas, the users of the highway will be primarily local users who use the highway on a daily basis for a certain number of trips. Buttlar et al. (2015) considered 12,000 miles as an average car travel in a year and multiplied the number of cars (AADT) with 12,000 miles operating costs. Then compared the operating costs with one mile of agency costs.

This approach is questionable because each car is assumed to travel to 30 miles approximately each day on a mile highway. Apparently, the operating expenses are multifold than the agency costs. To alleviate this problem, in this study, the operating costs were estimated by assuming a triangular probability distribution of cars traveled per day as 5 miles minimum to 30 miles maximum. But trucks have to travel only 5 miles a day (actual length of the highway). This assessment considers cars per day may have one to six trips on the five-mile highway. The IRI of various designs predicted using Pavement ME design software, and corresponding operating costs are shown in Fig. 6. Design 4 produced lowest operating costs as IRI is minimum than other designs.

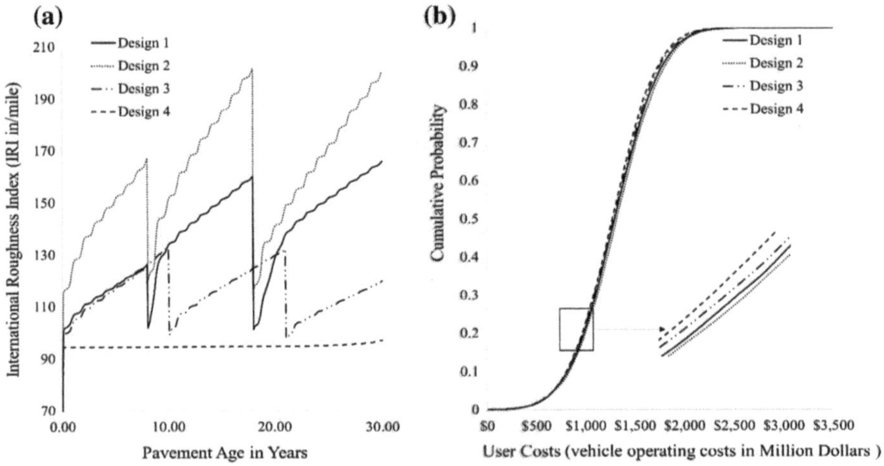

Fig. 6. **a** Pavements IRI. **b** Vehicle operating costs

7 Results and Discussions

Figure 7 displays the summary of various costs estimated in this study. For discussion purposes, it is assumed that NBWs are constructed to reduce the noise at 200 ft from highway and traffic delay costs are for 9AM–5PM maintenance hours. If only agency costs are considered for design selection, then Design 3 is the most appropriate. If the agency also focuses on reducing noise on neighborhood Design 2 is generating lower noise, but it needs regular maintenance which increases the agency maintenance costs. Design 4 requires a high initial construction cost, but it needs no major maintenance in 30 years as per the design software, and subsequently, the traffic delay costs are minimal. Out of all the costs, vehicle operating costs outplay the others. Design 4 outranks other designs in this category. It is evident that the rankings of designs are varying based on the priorities set. The authors believe including some other social impacts will shift the design rankings. This assessment shows the complexity in sustainable decision making. Hence, there is a need for intervention of experts in decision making with a robust and sustainable guideline.

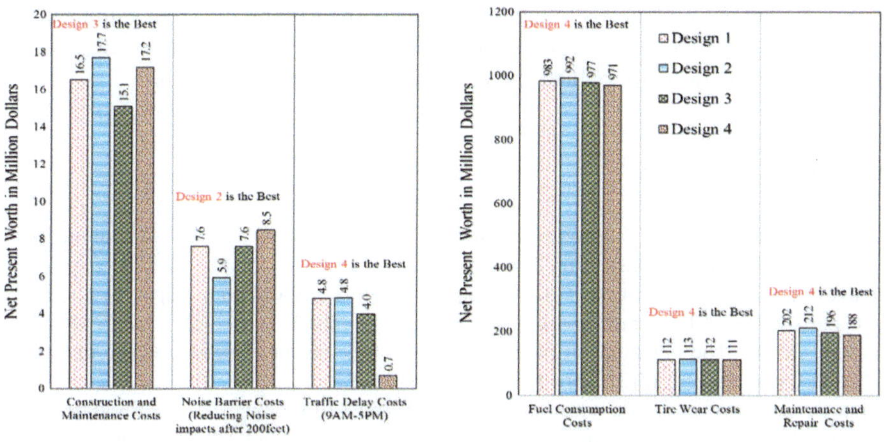

Fig. 7. Cost comparison of various pavement design alternatives

8 Summary

The main idea of this study was to show the influence of including social sustainability during design selection process. Designers can reduce some impacts on society by choosing right pavement structure design. An SLCA framework is proposed in this study to estimate impacts due to pavement design on people. The social impact categories and indicators selected in this study are just a subset and not necessarily a complete study. This research is not to explicitly state that design is a most sustainable one. The study only aims to shed light on various impacts that are solely based on selected pavement structure. In this study, four costs (noise barrier costs, fuel consumption, tire wear, and maintenance and repair costs) were identified and measured based on each pavement structure performance. This study employed probabilistic or risk approach using Monte Carlo Simulation for estimating all the costs. This method provided costs by considering the range of each key input. Out of the four designs, Design 3 has lower agency costs (construction of pavement and noise barrier costs), and Design 4 has the lowest user costs. The final decision on design will depend on the importance assigned to user costs by the state agency or transportation officials.

9 Limitations and Future Research

TNM does not consider the deterioration of pavement surface performance with time. Future research is required to evaluate the inconvenience to drivers because of noise reflection due to NBWs. The IRI values were calculated using AAHSTOWARE software, but the values need to be validated using the historical data using the state databases. Especially Design 4 has almost a constant IRI, which needs to be verified. The rolling resistance which is vital for estimating operating costs are calculated using IRI, but other factors like macro-texture and structural responsiveness need to include in the future study. In estimating operating costs, the impact of climate input was ignored. Future research needs to incorporate the climate data into costs estimates. This study did not consider the environmental costs (like GWP, Acidification, etc.), a future study needs to be done to include the environmental impacts as well.

References

Anderson, J., Muench, S.: Sustainability trends measured by the greenroads rating system. Transp. Res. Rec. J. Transp. Res. Board **2357**, 24–32 (2013)

Average Low Bid Unit Prices: Texas Department of Transportation, http://www.txdot.gov/business/letting-bids/average-low-bid-unit-prices.html (2016). Accessed 1 July 2016

Benoît, C., et al.: The guidelines for social life cycle assessment of products: just in time! Int. J. Life Cycle Assess. **15**(2), 156–163 (2010)

Benoît, C. (ed.): Guidelines for social life cycle assessment of products. UNEP/Earthprint (2010)

Bernhard, R., et al.: An introduction to tire/pavement noise of asphalt pavement. Institute of Safe, Quiet and Durable Highways, Purdue University (2005)

Buttlar, W.G., et al.: Integration of pavement cracking prediction model with asset management and vehicle-infrastructure interaction models (No. NEXTRANS Project No. 073IY03). US Department of Transportation, Washington, DC, USA (2015)

Chatti, K., Zaabar I.: Estimating the effects of pavement condition on vehicle operating costs, vol 720. Transportation Research Board (2012)

Consumer Price Index (CPI): U.S. http://www.bls.gov/cpi/ (2016). Accessed 1 July 2016

Cucurachi, S., et al.: A framework for deciding on the inclusion of emerging impacts in life cycle impact assessment. J. Clean. Prod. **78**, 152–163 (2014)

Dubois, A., Gadde, L.E.: The construction industry as a loosely coupled system: implications for productivity and innovation. Constr. Manage. Econ. **20**(7), 621–631 (2002)

Federal Highway Administration (FHWA) Sustainable highways initiative overview. https://www.sustainablehighways.dot.gov/overview.aspx#quest2 (2015). Accessed 21 July 2015

Fenandez-Solis, J.L.: How the construction industry does differ from manufacturing

Gidado, K.I.: Project complexity: the focal point of construction production planning. Constr. Manage. Econ. **14**(3), 213–225 (1996)

Grießhammer, R., et al.: Feasibility study: integration of social aspects into LCA. Öko-Inst, Freibg. (2006)

Hunkeler, D.: Societal LCA methodology and case study (12 pp). Int. J. Life Cycle Assess. **11**(6), 371–382 (2006)

Lew, J.B.: Informing roadway sustainability practices by using greenroads certified project data. Transp. Res. Rec. J. Transp. Res. Board (2589), 1–13, Transportation Research Board of the National Academies, Washington, D.C. (2016)

Muzet, A.: Environmental noise, sleep and health. Sleep Med. Rev. **11**(2), 135–142 (2007)

NCHRP Report 720: Estimating the effects of pavement condition on vehicle operating costs. Transportation Research Board of the National Academies (2012)

Noise and Hearing Conservation Technical Manual Chapter: Appendix I:C. Effects of excessive exposure. https://www.osha.gov/dts/osta/otm/noise/health_effects/effects.html#revised. Accessed 8 July 2015

Olander, S.: External stakeholder analysis in construction project management. Lund University (2006)

Paragahawewa, U., et al.: Social life cycle analysis (S-LCA): some methodological issues and potential application to cheese production in New Zealand. Report by Agresearch (2009)

Rangaraju, P.R., et al.: Life cycle cost analysis for pavement type selection (No. FHWA-SC-08-01). South Carolina Department of Transportation (2008)

R.S. Means Company, Incorporated: RSMeans heavy construction cost data (2012)

Russel, L.W.: Pavement design guide. Texas Department of Transportation (2011)

Social Hotspots Database: http://socialhotspot.org/ (2014). Accessed 21 July 2014

Schmidt, I., et al.: SEEbalance®: managing sustainability of products and processes with the socio-eco-efficiency analysis by BASF. Greener Manage. Int. **45**, 79 (2004)

Spangenberg, J.H.: Design for sustainability (DfS): interface of sustainable production and consumption. Handbook of Sustainable Engineering, pp. 575–595. Springer, Netherlands (2013)

Thomas, J., et al.: Towards sustainable pavement systems: a reference document. Federal Highway Administration (2015)

U.S. Department of Transportation: Life cycle cost analysis primer (2002)

Weidema, B.P.: ISO 14044 also applies to social LCA. Int. J. Life Cycle Assess. **10**(6), 381 (2005)

A Cost-Effective Approach Towards Road Construction—Kondave a Case Study

Anuj Gade[1], Sushma Kulkarni[1], Anand Tapase[2(✉)], and Sanjiv Bonde[2]

[1] Rajarambapu Institute of Technology Rajaramnagar, Islampur, India
anujgade.ag@gmail.com, sushma.kulkarni@ritindia.edu
[2] Karmaveer Bhaurao Patil College of Engineering, Satara, India
{tapaseanand, sanjivbonde}@gmail.com

Abstract. The planning, design, and construction of roads in India are depended on the local traffic requirements and the recommended design traffic in terms of a cumulative number of standard axles. The roads are getting deteriorated in the form of potholes, fatigue cracking, rutting deformation as a number of heavy, overloaded, high tyre pressure vehicles are using the ordinary/village roads which are affecting the riding comfort of vehicles resulting into fatal injuries to road users. Investigations in India and abroad have revealed that the abundantly available waste materials like fly ash, rubber, waste plastic, e-waste, etc. which are creating disposal problem can be effectively used in road construction. A small village Kondave on the State Highway 141 joining the Satara district to the world famous hill station Mahabaleshwar on Sahyadri ranges was considered as a case study. The tremendously increasing visitors using this route are throwing used plastic carry bags, bottles, etc. on a large scale creating its disposal problem in front of local people. The present work deals with the assessment of the generation of plastic waste in the village along with to check its suitability in road construction. In the initial stage, the available plastic waste from the village was collected, segregated, recycled and grinded. To check its suitability in road construction a thorough experimental investigation was carried out using a Marshall Stability procedure. The mix were prepared as per SP-98-2013 for a dry process wherein the poor locally available aggregates are coated with waste plastic and as per SP-53-2010 for a wet process wherein the bitumen is modified with its partial replacement. The results reveal that the Marshall Stability value of poor conventional materials can be improved by the application of 7% optimum waste plastic. The application of waste plastic in the road not only mitigates the disposal problem but shown positive improvements in strength and other governing parameters including viscosity. So, instead of using a higher bituminous mix in the top layer for limiting the top-down cracking and bottom-up cracking modified bitumen is a cost-effective alternative. Check on construction cost is achieved, as the poor locally available materials are made in use due to the application of waste plastic. Hence, such type of cost-effective approach for the construction of roads along with eco-friendly disposal of plastic waste can be used to mitigate the early deterioration of roads.

Keywords: Flexible pavement · Cost effective

W. Steyn et al. (eds.), *Pavement Materials and Associated Geotechnical Aspects of Civil Infrastructures*, Sustainable Civil Infrastructures,
https://doi.org/10.1007/978-3-319-95759-3_8

1 Introduction

Due to upgradation and development plans, the number of heavy, overloaded, high tyre pressure vehicles are more frequently using the ordinary/village roads which are not designed for such vehicles. As a consequence, the roads are getting deteriorated in the form of potholes, fatigue cracking, rutting deformation which are affecting the riding comfort of vehicles resulting in fatal injuries to road users. To mitigate the number of accidents caused due to early deterioration of roads and increased maintenance cost, concentration should be focused on the use of alternative more suitable materials to the conventional materials along with the design procedures. The failure of roads in the form of top-down cracking or the conventional bottom-up cracking is the result of excessive tensile stresses developed at the top of the surface due to such type of excessive loads and tyre pressures. The investigation has revealed that the modern technology and the increasing demand of industry have resulted in the major loss of environment and the mankind. Nondegradable manmade material soon after its beneficial use to the society results in creating nuisance during its disposal. Dumping of a nonrecyclable, nondegradable waste material is the thrust area of todays research as ultimately a socio-friendly material affects the ecological balance. On the other hand, the conventional materials fail in sustaining the drastic effect of modern vehicles and its overloading, which results in deterioration of roads with a constant pace in the form of rutting and fatigue.

Every fundamental sector starting from agriculture to packaging is almost revolutionized by the applications of plastic/polymer in various forms due to its multipurpose use. Plastic is a non-biodegradable material and several studies have proven the health hazard like reproductive problems in human and animal, genital abnormalities etc. caused due to its improper disposal. Researchers have already reported that the material can remain for a long duration on earth without degradation. A complete ban on plastic may hammered the various sectors in heavy financial losses and looking forward to the present day lifestyle it seems to be an impossible task. But due to its non-biodegradable nature, the hips of waste or used plastic is taking the face of a devil for the present and future generation. So the present study deals with solving the disposal problem by reusing the plastic waste and finding an eco-friendly alternative for its disposal leading to sustainability.

Investigations in India and abroad have revealed that the abundantly available waste materials like fly ash, rubber, waste plastic, e-waste, etc. which are creating disposal problem can be effectively used in road construction. In the present study, the problems due to the disposal of plastic waste are identified and are studied at length experimentally along with its application in a form of constructing a road patch. The objective of work is to investigate the effect of plastic waste in the flexible pavement and to suggest the optimum percentage of bitumen that can be replaced by plastic waste for the improvement of roads. Also, the study revealed the use of sustainable alternative material over conventional materials with cost reduction and improvement in various pavement governing parameters.

2 Literature Review

Researchers world over have investigated experimentally on the use of low-density polyethylene (LDPE) and high-density polyethylene (HDPE) and obtained positive results which have preached a path for the application of plastic in road construction. Low-density polyethylene (LDPE) and High-density polyethylene (HDPE) are been used before in experimental works related to flexible pavements showing positive results related to use of plastic waste in flexible pavements. The polymer coated aggregate bitumen mix perform better for flexible pavements**1**. The results for both dry and wet specimen of 8% waste polymer modified bitumen (WPMB) mix show 50% enhanced tensile strength compared to conventional mix and more resistance to water damage. Punith and Veeraragavan used reclaimed PE as asphalt modifiers. They found that the basic test properties indicated that the addition of PE content to neat asphalt reduces the penetration and ductility values, whereas increases in softening point and specific gravity values were observed with the addition of PE modifier. A PE content of 5% by weight of asphalt is recommended for the improvement of the performance of asphalt cement. Sinan and Emine tried to investigate the possibility.

3 Case Study

A small village Kondave on the State Highway 141 joining the Satara district to the world famous hill station Mahabaleshwar on Sahyadri ranges was considered as a case study. Due to growing popularity number of development plans are ongoing which has given rise to heavy commercial vehicles. The tremendously increasing visitors using this route are throwing used plastic carry bags, bottles, etc. on a large scale creating its disposal problem in front of local people. It was decided to make survey related to the plastic waste generation and its disposal in Kondave village The population of Kondave village was 2500 in 2001; the Current population of Kondave village is 8900 (As per the data are given by Grampanchyat office Kondave village). There are five solid waste depots in Kondave village. Solid waste is collected by vehicle and placed in the solid waste depot. In Kondave village waste plastic is disposed by landfilling or incineration but those methods are dangerous to human health and environment. The main aim of the site selection of Kondave village is disposal plastic waste in eco-friendly way and application in the construction of flexible pavement. Kondave village lies on State Highway 141 (Vijapur—Rahimatpur—Satara—Mahabaleshwar) which is located at a distance of 4 km away from Satara city and 45 km from Mahabaleshwar. It has a population of around 8900 people. In the village, a common garbage dumping ground is available wherein all type of household and related waste products including plastic are thrown.

4 Methodology

To check the suitability of plastic in road construction a thorough experimental investigation was carried out using a Marshall Stability procedure. The mix was prepared as per SP-98-2013 for a dry process wherein the poor locally available aggregates are coated with waste plastic and as per SP-53-2010 for a wet process wherein the bitumen is modified with its partial replacement. The work consists of an experimental approach towards cost-effective waste management and finding an alternative to conventional materials in flexible pavements. The results obtained in laboratory investigation indicate a major gain in strength with a substantial saving in cost. A number of laboratory tests have been carried out partially replacing bitumen by plastic waste.

A. *Bitumen*

Bitumen is defined as "A solid or viscous liquid, consisting essentially hydrocarbons and their derivatives, which is soluble in trichloroethylene and is substantially non-volatile and softens gradually when heated. Bitumen is collected from hot mix bitumen plant which is nearer to Kondave village for experimental purpose. The bitumen used for the experiment was 60/70 grade and was tested in the laboratory for basic tests, penetration, ductility, softening point, flash and fire point etc.

B. *Aggregate*

Aggregates are natural materials and are the most important component of road construction Aggregates are also used as base material under road construction. Aggregate is collected from the stone crusher, which is nearer to Kondave village. The aggregate was tested in the laboratory for crushing strength, impact test, specific gravity, Los Angeles abrasion value etc. Aggregates are divided into two categories according to size.

C. *Plastic*

Plastic is a material consisting of wide range of organic compounds (synthetic or semi-synthetic) that are malleable and can be molded into solid objects. Plastics are typically organic polymers of high molecular mass, but they often contain other substances. Safe disposal of waste plastic is a serious environmental problem. Being a non-biodegradable material it does not decay over time and even if dumped in landfills, finds its way back into the environment through air and water erosion can choke the drains and drainage channels.

Waste plastic contains milk pouch, packaging bags, plastic bottles, protective plastic packaging, sacks, milk pouches, household articles, carry bags, plastic film, straws, Plumbing pipes, cosmetic bottles, industrial plastic covers etc. Normally plastic is classified into thermosetting and thermoplastic.

D. *Modified Bitumen*

Plastic waste is mixed in bitumen by wet mix process. Modified bitumen is a partial replacement of bitumen by waste plastic [5]. Replacement quantity of waste plastic is 6.5% of bitumen weight as per IRC: SP: 53-2010 [3].

E. *Plastic Coated Aggregate*

Plastic waste is mixed in aggregate by dry mix process. This method carried out by using central mix plant. For laboratory experiment, the aggregate is heated up to 165 ° C then blending of shredded waste plastic in hot aggregate [4]. The quantity of waste plastic for the dry process is percent value (As per IRC: SP: 98: 2013) of optimum bitumen content. In dry process plastic is coated over the surface of the aggregate is known as a Plastic coated aggregate.

F. *Mix Design*

The gradations adopted for mix are taken as specified by MORTH 2013 (Table 500-2, PP-151). Mix design is made of OGC and Seal coat layer of flexible pavement as shown in the table, in which aggregates size of 12.5, 10, 2.6 mm and grit is used [5]. Steps carried while preparing the mix are as;

(a) Oven dried aggregate for approximately 12 h at around 105–110 °C was separated into the individual specified sizes by dry sieving. (b) Individual aggregates as per the adopted gradations were recombined to the correct proportions. (c) The aggregates were thoroughly mixed on mixing pan. (d) Then the hot bitumen as per trial percent is mixed with aggregate. (e) After properly mixing aggregate with bitumen the mix was filled with a mould at a time and compaction plate was placed on which 75 number of blows were given on either side of sample manual by Marshall hammer.

For replacement of plastic as per the considered trials before proceeding to step (d) shredded plastic is sprinkled on hot aggregate and then conducting step (d) for dry processes or mixing shredded plastic with hot bitumen for wet process (Tables 1 and 2).

Table 1. Types and size of plastic

Name of plastic	Type of plastic	Size (μ)
Milk pouch	LDPE	60
Parcel cover	LDPE, PE	50
Carry bags	LDPE, PE	10
Water bottle	HDPE	210
Biscuit cover	Polyester + PE	40
Packing cover	LDPE	50

5 Result and Discussion

All tests on bitumen and aggregate are conducted in Transportation Engineering Laboratory of RIT College, Islampur. Bitumen is replaced by waste plastic in percentage from 6.5 to 7.5% by weight of bitumen (As per IRC: SP: 98-2013). A property of bitumen and aggregate is improved after mixing of 7% of waste plastic by a dry process and 6.5% of waste plastic by a wet process. Wet mix process is used for mixing of waste plastic in bitumen. Here the specimens are prepared as per standard Marshall Stability test (ASTM D 1559) and Tables 3 and 4 shows the obtained results of various trials by both the methods. Replacement of bitumen by plastic is economical Rs 321

Table 2. Mix design

Quantity for 10 m^2		
Material	Quantity	Percentage value
For OGC		
(a) Nominal stone size 13.2 mm	0.18 m^3	
(b) Nominal stone size 11.2 mm	0.09 m^3	
(c) Bitumen	14.6 kg	
For seal coat		
(a) Nominal stone size shall pass 2.36 mm sieve and retained 180-μ sieve mm	0.09 m^3	
(b) Bitumen	9.8 kg	
Total quantity		
(a) Nominal stone size 13.2 mm	0.18 m^3	52.00
(b) Nominal stone size 11.2 mm	0.09 m^3	23.92
(c) Nominal stone size shall passing 2.36 mm sieve and retained 180-μ sieve mm	0.09 m^3	23.92
(d) Bitumen	24.40 kg	4.33

Table 3. Result of Marshall stability test by dry process

S. no.	Weight of sample (gm)	Height (cm)	Bulk density (gm/cc)	Stability (KN)	Flow (mm)	% Air voids
Control mix						
1	1190	6.33	2.31	10.70	3.52	5.21
2	1218	6.37	2.34	10.50	3.48	
3	1185	6.50	2.20	10.20	3.50	
Average			2.30	10.46	3.50	
6.5% Bitumen replaced by plastic						
1	1225	6.30	2.38	11.65	3.54	3.86
2	1195	6.36	2.31	11.58	3.52	
3	1233	6.26	2.30	11.48	3.50	
Average			2.33	11.50	3.52	
7% Bitumen replaced by plastic						
1	1205	6.23	2.32	11.42	3.52	3.40
2	1223	6.30	2.37	11.65	3.56	
3	1213	6.33	2.34	11.55	3.51	
Average			2.34	11.54	3.53	
7.5% Bitumen replaced by plastic						
1	1220	6.26	2.38	11.57	3.54	3.86
2	1226	6.33	2.32	11.52	3.52	
3	1230	6.30	2.29	11.48	3.52	
Average			2.33	11.52	3.52	

Table 4. Result of Marshall stability test by wet process

S. no.	Weight of sample (gm)	Height (cm)	Bulk density (gm/cc)	Stability (KN)	Flow (mm)	% Air voids
Control mix						
1	1190	6.33	2.31	10.70	3.52	5.20
2	1218	6.37	2.34	10.50	3.48	
3	1185	6.50	2.20	10.20	3.50	
Average			2.30	10.46	3.50	
6% Bitumen replaced by plastic						
1	1205	6.30	2.32	11.05	3.53	4.54
2	1210	6.22	2.30	10.70	3.50	
3	1195	6.26	2.31	11.25	3.49	
Average			2.31	11.00	3.51	
6.5% Bitumen replaced by plastic						
1	1205	6.24	2.27	10.60	3.52	5.20
2	1200	6.20	2.28	10.55	3.56	
3	1213	6.33	2.34	11.10	3.51	
Average			2.30	10.75	3.53	
7% Bitumen replaced by plastic						
1	1220	6.26	2.34	11.15	3.54	5.20
2	1195	6.35	2.27	10.60	3.52	
3	1210	6.40	2.29	10.65	3.52	
Average			2.30	11.80	3.52	

saved by the use of waste plastic in the flexible pavement. In general study, the cost for collection of plastic is considered up to Rs 10. After considering the cost for collection of waste plastic, this method is also economical. This method is 12% economical as compared to normal control mix (Figs. 1 and 2).

Fig. 1. Current status of waste plastic depot

Fig. 2. Wet mix process at laboratory

6 Conclusion

Based on the experimental study following important conclusions are drawn.

(1) The results obtained in laboratory investigation indicates a major gain in strength required for road construction with substantial cost saving.
(2) Total plastic waste quantity generated at Kondave village is 330 kg/month; it includes house plastic waste and industrial plastic waste. The plastic waste contains maximum quantity of low-density polyethylene and high density of polyethylene. Low-density polyethylene is more feasible in bitumen as compare to high-density polyethylene. As per survey 80% waste plastic from LDPE in Kondave village.
(3) In the wet mix process for modified bitumen testing 6.5% accurate percentage value for replacement of bitumen by waste plastic. Penetration value is decreased by 6.68% after mixing of 6.5% waste plastic in bitumen but the softening point of bitumen is increased by 8.60% and other properties of bitumen are same. In Dry mix process for aggregate testing, 7% waste plastic quantity is accurate as compared to another percentage variation. 7% optimum bitumen content is replaced by plastic waste; it forms the plastic coated aggregate. The specific gravity of plastic coated aggregate is increased by 2.88% after coating of 7% waste plastic on aggregate also crushing value, impact value, loss abrasion value is decreased by 3–4%. This all result of the aggregate test indicates that replacement of bitumen by plastic waste is increasing the properties of aggregate. Plastic coated aggregate increase the stones improving the surface property of aggregates. Plastic coating form thin layer around the aggregate to fulfill parameter and binding property of aggregate. Plastic coated aggregate is used for the good performance of the flexible pavement.
(4) Marshall Stability value of bituminous material increased in dry mix process and wet mix process. In wet mix process, 6% plastic content is optimum shows the better result as compare to other content. In Dry mix process plastic content is 7%,

shows high Marshall stability value. Due to the addition of waste plastic, Marshall Value increase as compare to control mix. High Marshall values show high strength, high durability, high load carrying capacity. This all parameter shows increase the properties of flexible pavement.

(5) As compare to wet mix process, dry mix process is suitable method for mixing of waste plastic in bitumen for construction of flexible pavement because wet mix process required more time and energy for blending; new equipment are required for wet mix process, Dry mix process is simple and feasible.

References

1. Al-Hadidy, A.I., Yi-Qui, Tan: Effect of polythene on the life of flexible pavements. Constr. Build. Mater. **12**, 1456–1464 (2009)
2. IRC SP-53:2010: Guidelines on use of modified bitumen in road construction
3. IRC: SP: 98-2013: Guidelines for the use of waste plastic in hot bituminous mixes (DRY PROCESS) in wearing courses
4. Ministry of Road Transport and Highways: Specifications for Road and Bridge Works, fifth Revision. Indian Road Congress ISI 13300 Part III (2013)
5. Punith, V.S., Veeraragavan, A.: Behaviour of asphalt concrete mixtures with reclaimed polyethylene as additive. J. Mater. Civil Eng. American Society of Civil Engineers, USA, pp. 33–44 (2007)
6. Ranadive, M.S., Tapase, A.B.: Improvement in strength of flexible pavement: an experimental approach. J. Environ. Res. Dev. **6**(3), 28–32 (2013)
7. Angelone, S., Cauhape, M., Manuel, C., Fernando, B., Martinez, O.: Green pavements: reuse of plastic waste in asphalt mixtures. Mater. Struct. **49**, 4–8 (2016)

Geophysical Properties of Sand-Cement-Inorganic Binder Mixture: Electrical Resistivity and Elastic Wave Velocity

Pacifique Kiza Rusati, Sanha Kim, and Ki-Il Song(✉)

Department of Civil Engineering, Inha University, 100, Inha-ro, Nam-gu, Incheon, Republic of Korea
ksong@inha.ac.kr

Abstract. This paper investigates the effects of mixing ratio, cement content and curing time on the transient electrical and mechanical properties of sand-cement-inorganic binder mixture adopting two non-destructive testing (NDT) methods, electrical resistivity (ER) test and Free Free resonant Column test (FFRC). Due to the environmental issues related to cement, the reduction of cement usage is requesting. Inorganic binder is then introduced in this paper as alternative to reduce the use of cement contents for ground improvement. The results reveal that mixing ratio and curing time have considerable effects on electrical and mechanical properties of the mixture, and it is expected that sufficient strength can be obtained with the increase of inorganic binder content, confirming that the usage of cement can be reduced to some extent.

1 Introduction

Cement grout, also known as cement milk grout, in which cement is used as principal binder, is a common type of grouting materials used for ground improvement. However, cement is harmful to the environment from its production to its use in construction (Mavroulidou et al. 2013; Ogedengbe and Oke 2011; Estevez et al. 2003). Furthermore, construction difficulties such as bleeding and leaching are also related to cement use in construction. These problems can be overcome by reducing utilization of cement and using alternative eco-friendly binders. Inorganic binders are particularly cost-effective and environment-friendly; also they are capable to sustain the strength of grout material (Azadi et al. 2017; Warner 2004), and enhance grout performance when they partially or fully replace cement in the mixture (Mavroulidou et al. 2013; Güllü et al. 2017; Toma et al. 2011).

Quality control of grout is one of the major stage to be considered during grouting. However, conventional testing methods, such as compression and shear tests, used to estimate properties of materials, present the disadvantage of creating deformation in the material and altering its properties (Das 2013); thus they may not be effectively applied in the field. Therefore, non-destructive-based elastic and electromagnetic wave methods are nowadays widely being adopted to evaluate geomaterials' characteristics since wave parameters can be directly and indirectly correlated with several geophysical properties (Herrick and kennedy 1994; Verástegui-Flores et al. 2015).

W. Steyn et al. (eds.), *Pavement Materials and Associated Geotechnical Aspects of Civil Infrastructures*, Sustainable Civil Infrastructures,
https://doi.org/10.1007/978-3-319-95759-3_9

In this study, electrical resistivity, longitudinal wave velocity and damping ratio of sand-cement mixtures improved with addition of inorganic binders are measured through Electrical resistivity (ER) test and Free Free resonant Column (FFRC) test, and effects of mixing ratio, cement content and curing time on the transient electrical and mechanical properties of sand-cement-inorganic binder mixtures are evaluated.

2 Experimental Tests and Materials

Experimental tests
ER test measures the current flowing through different horizontal layers composing the grout sample (Rhett 2001). ER test configuration is illustrated in Fig. 1. Electrical resistivity of grout can be estimated using Eq. (1).

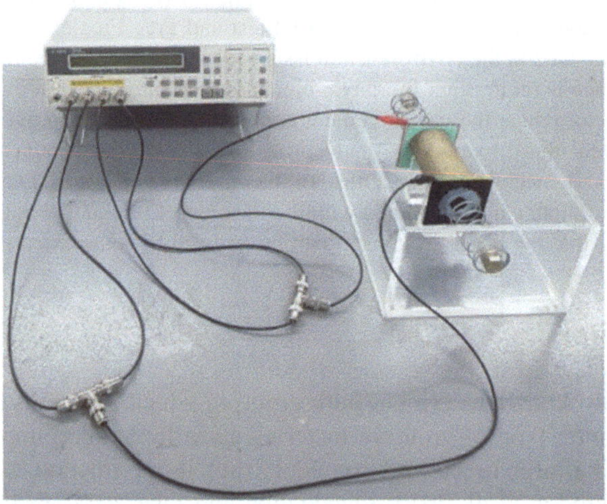

Fig. 1. ER test apparatus configuration

$$\rho = R\frac{A}{L} \tag{1}$$

where ρ is the sample electrical resistivity (unit: Ω-m), R is the resistance(unit: Ω), and *A* and *L* being respectively sample cross-sectional area in m^2 and length in m). The resistance is the ratio of voltage V in volt and current I in ampere ($R = \frac{V}{I}$).

ER test was performed using a set up composed of an LCR meter (Agilent 4263B) connected to two clips Kelvin leads, each paired with two leads wires inserted to current and potential terminals of the LCR meter, and the support apparatus (Fig. 1). Sample was placed on two thin acrylic layers and two copper meshes was attached to both its cross-sectional area by two spring in the support apparatus. The clips were then connected to the copper meshes and transmit information to the LCR meter, which measures the sample electrical resistance. The recorded resistance is then used to calculate electrical resistivity of the sample according to Eq. (1).

FFRC test is a simple non-destructive testing method which uses stress wave (Sansalone and Streett 1997; Azari et al. 2014). To carry out FFRC test, test arrangement up composed of an oscilloscope (Agilent DSO-X-3024A), a signal conditioner (PCB 480B21), an accelerometer (PCB 352C22) and a mechanical hammer was established. As described in Fig. 2, grout sample was suspended in air with two thin wires to facilitate free boundary conditions, and accelerometer was installed to the right end of the sample in axial direction with the point of impact. The hammer was used to impact the left end of the sample and displacement response of longitudinal wave due to that impact is recorded in time-domain waveform and transformed to frequency-domain (Fig. 3) through fast Fourier Transform (FFT). From obtained frequency response, longitudinal wave velocity (V_L) and geometric damping ratio of sample can be calculated using Eqs. (2) and (3).

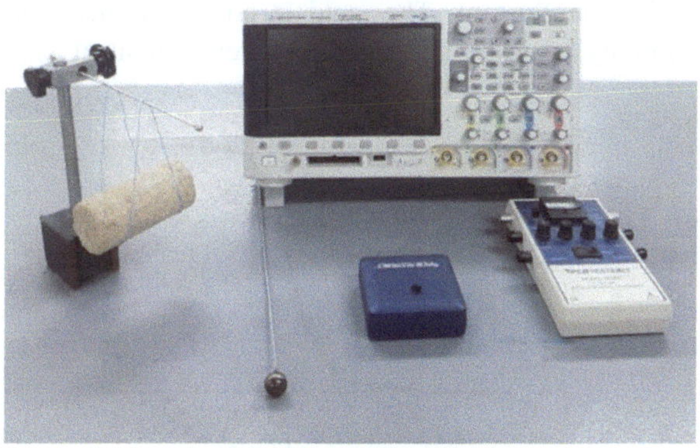

Fig. 2. FFRC test arrangement

$$V_L = 2Lf_n \tag{2}$$

$$H(f) = \frac{1}{\left[1 + J2D_G\left(\frac{f}{f_n}\right) - \left(\frac{f}{f_n}\right)^2\right]} \tag{3}$$

where L is the height of sample (in m), V_L is longitudinal wave velocity (in m/s), f_n is natural frequency (in Hz), *H(f)* is transfer function and J is an imagery index (Azari et al. 2014; Song and Cho 2009; Mustaqim et al. 2014).

Testing materials

Grout samples used for this experiment were prepared using sand, Portland cement and binding materials.

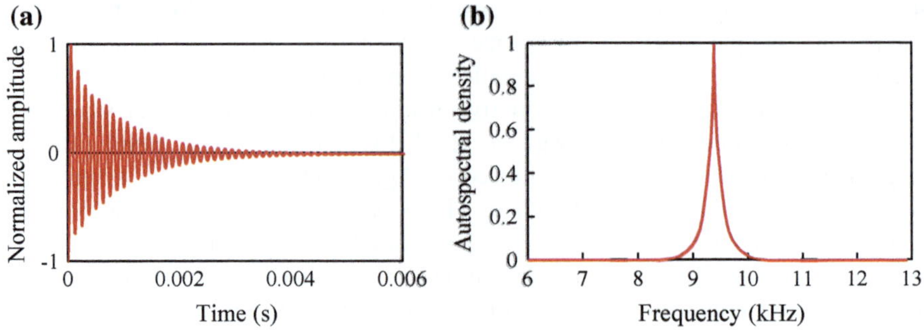

Fig. 3. Typical result from FFRC test. **a** Time domain signal. **b** Frequency response

Moisture content test and Sieve analysis were carried out in laboratory to characterize the particle size distribution, specific gravity, and water content. Grain size distribution (Fig. 4) is performed based on ASTM D2487 and classified according to Unified Soil Classification System. The sand is coarse-grained poorly graded sand (SP) with specific gravity of 2.66 and moisture water content of 2.83.

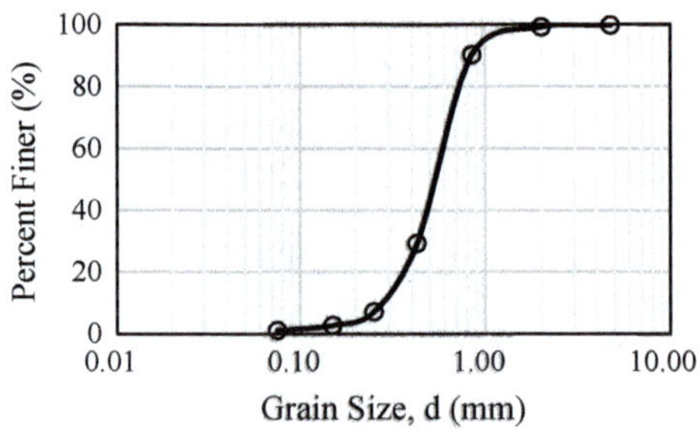

Fig. 4. Grain size distribution curve of sand used for experiment

Portland cement Type I was used for this experiment in accordance with ASTM C150 and its physical properties are grade 32.5 MPa, specific gravity: 3.12 g/cm^3 and fineness 1.77.

Two inorganic binders, S100 and S200 were added to the sand-cement mixture. They are composed of Calcium oxide CaO, Aluminium oxide Al_2O_3 and calcium sulfate $CaSO_4$ as reactants in a saturated state. S100 is a liquid binder reacting as pozzolanic material and generating large amount of strong alkaline materials ($Ca(OH)_2$) during hydration of cement which allows the grout to gain the required strength (Motamedi et al. 2015; Azadi et al. 2017), and prevent contamination due to water leachate. When S200 is mixed in cementitious materials, its main component, calcium aluminate $CaAl_2O_4$ reacts with water to produce various aluminates; particularly in the presence of SO_3, monosulfate ($3CaO.Al_2O.3CaSO_4.12H_2O$) is produced and calcium silicate hydrate (C–S–H) is obtained.

108 samples of sand-cement-inorganic binder mixture were prepared in the laboratory and cured for 7, 14 and 28 days. Samples were cylindrically shaped, each with 5 cm diameter and 10 cm height. Tests were performed while varying the mixing ratio of water volume, cement content and inorganic binders' content in a constant sand quantity in order to evaluate effects of mixing ratio on the mixture. Types A and B were prepared separately before being mixed with sand. The contents of samples are tabulated in Table 1.

Table 1. Mixing ratio and reduction of cement for sand-cement—Inorganic binder mixture

Mixing ratio	Type A					Type B		Sand	
	S200 (mg)	Cement content (mg)			Water (ml)	S100 (ml)	Water (ml)	Content (mg)	%
		0%	−10%	−20%					
C-150	1.62	27.83	25.05	22.27	25.62	0.25	3.68	355.3	76.9
C-200	2.16	37.11	33.40	29.69	34.16	0.33	4.91	353.5	71.4
C-300	3.24	55.67	50.10	44.53	51.25	0.50	7.36	353.3	62.4
C-350	3.78	64.94	58.06	51.95	59.79	0.59	8.59	353.3	58.8

3 Results and Discussion

Effects of Mixing Ratio, Cement Content and Curing Time on Electrical Resistivity

Electrical resistivity of mixture decreases with cement content decrease and mixing ratio increase (Fig. 5a). Due to water high conductivity effects, the increasing water content when mixing ratio increase causes the decrease of electrical resistivity (Liu et al. 2008). Result also shows that the influence of cement content of sand-cement-inorganic binder mixture on electrical resistivity is negligible at high mixing ratio. Electrical resistivity of grout also increases with the increasing curing time (Fig. 6a).

Effects of Mixing Ratio, Cement Content and Curing Time on Longitudinal Wave Velocity

Results of FFRC test proved that longitudinal wave velocity decreases with the decrease of both cement content and mixing ratio (Fig. 5b). The increase of longitudinal wave velocities of high mixing ratio cases (C-300, C-350) is steeper than that of low mixing ratio cases (C-200, C150). This implies that reaction and hardening of admixture is relatively faster in high mixing ratio cases. Furthermore, longitudinal wave velocity displays an increasing tendency when grout is cured (Fig. 6b).

Effects of Mixing Ratio, Cement Content and Curing Time on Damping Ratio

Damping ratio of mixture decreases as the mixing ratio increases, but marginally increases with the decreasing of cement content (Fig. 5c). This is due to creation of more bonds between particles in the mixture (Saxena et al. (1987) generated by high hardening level caused by reaction of inorganic binders in the mixture. The trend of damping ratio change with respect of cement content is consistent with Saxena et al.

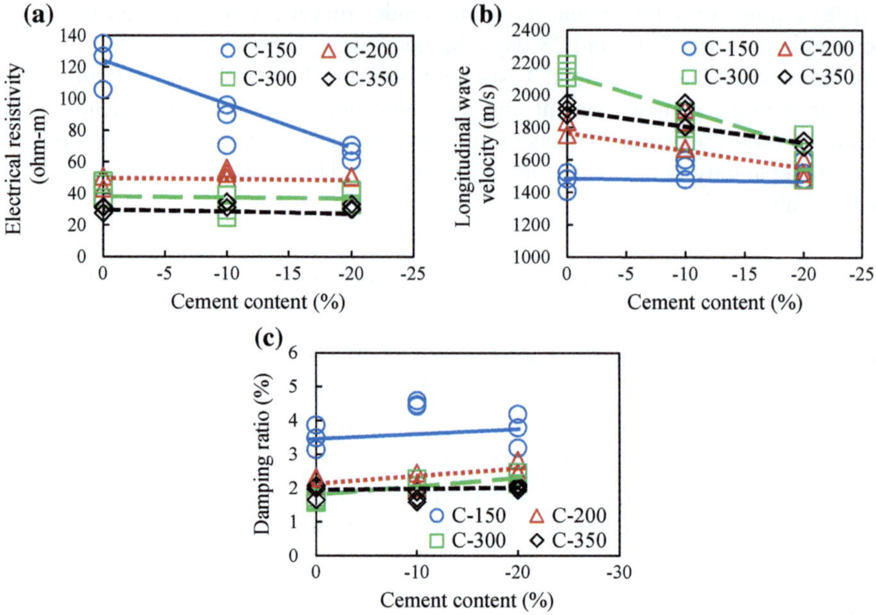

Fig. 5. Variations of **a** electrical resistivity, **b** longitudinal wave velocity and **c** damping ratio due to cement content reduction after 28 days curing period

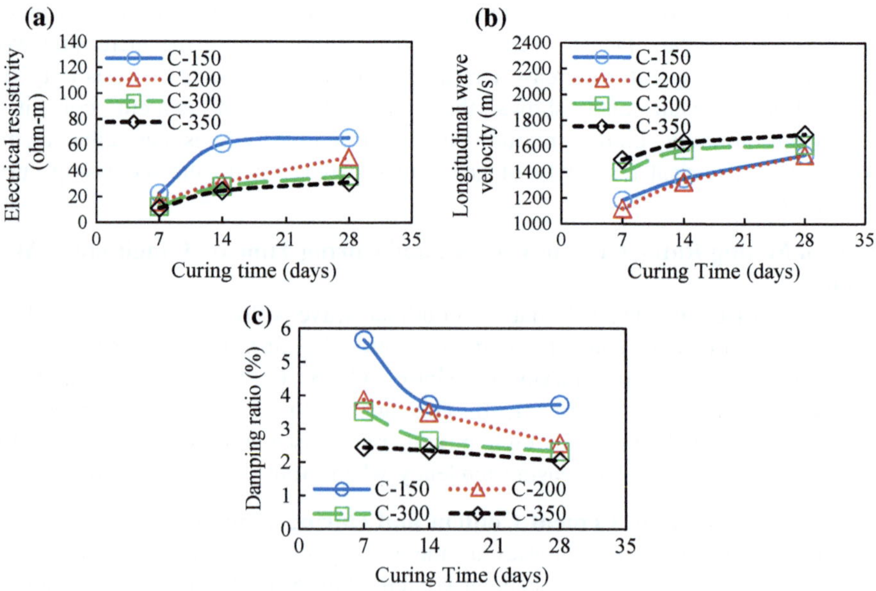

Fig. 6. Effects of curing time on **a** electrical resistivity, **b** longitudinal wave velocity and **c** damping ratio when 20% cement content is reduced

(1987) who observed the increase of damping ratio with reduction of cement content. From Fig. 6c, it is also observed that damping ratio decreases with the increase of curing time. However, considerable effects of cement content and curing time on mixture cannot be found when the mixing ratio is high (i.e. C350).

4 Discussion

Inorganic binders were added in the sand-cement mixture. However, trends from this experiment are still consistent with several previous studies, such as Liu et al. 2008 and Saxena et al. 1987, performed on mixture without inorganic binder. Furthermore, previous investigations carried out on mixture with inorganic binders have demonstrated that increasing inorganic binder content in the mixture resulted in the increase of the strength of the mixture (Toma et al. 2011; Mavroulidou et al. 2013; Motamedi et al. 2015; Azadi et al. 2017; Güllü et al. 2017). Therefore, it is expected that inorganic binder can be successfully used in the field and sustain strength when its content is increased in the mixture. Nevertheless, further researches are required to fully apply approach suggested in this study in the field.

5 Conclusion

In this study, it is observed that mixing ratio, cement content and curing time have considerable effects on electrical resistivity, longitudinal wave velocity and damping ratio of the mixture. When mixing ratio increases, longitudinal wave velocity increases but electrical resistivity and damping ratio decreases. As samples are being cured, electrical resistivity and longitudinal wave velocity increase but damping ratio decreases. Trends from this study demonstrate that geophysical characteristics of grouting materials can be successfully estimated in the field using proper NDT methods. Moreover, the approach proposed in this study may not be limited to sand-cement-inorganic binder mixture only, but can also be applied to characterize other geomaterials.

References

ASTM C150 A: Standard specification for portland cement. ASTM International, West Conshohocken, PA, USA (1999)

ASTM D2487: Standard Practice for classification of soils for engineering purposes. ASTM International, West Conshohocken, PA, USA (2006)

Azadi, M.R., Taghichian, A., Taheri, A.: Optimization of cement-based grouts using chemical additives. J. Rock Mech. Geotech. Eng. **2017**, 1–15 (2017)

Azari, H., Nazarian, S., Yuan, D.: Assessing sensitivity of impact echo and ultrasonic surface waves methods for nondestructive evaluation of concrete structures. Constr. Build. Mater. **71**, 384–391 (2014)

Das, B.M.: Fundamentals of geotechnical engineering, 4th edn. Engage Learning (2013)

Estevez, B., Aguado, A., Josa, A.: Environmental impact of concrete recycling, coming from construction and demolition waste (c&dw). Deconstruction and Materials Reuse CIB Publication. In: Proceedings of the 11th rinker international conference (2003)
Güllü, M., Canakci, M., Al Zangana, I.F.: Use of cement based grout with glass powder for deep mixing. Constr. Buildings Mater. **137**, 12–20 (2017)
Herrick, D.C., Kennedy, W.D.: Electrical efficiency-A pore geometric theory for interpreting the electrical properties of reservoir rocks. Geophysics **59**(6), 918–927 (1994)
Liu, S.Y., Du, Y.J., Han, L.M., Gu, M.F.: Experimental study on the electrical resistivity of soil–cement admixtures. Environ. Geol. **54**(6),1227–123 (2008)
Mavroulidou, M., Boulouki, G., Unsworth, C.: Incorporating waste paper sludge as partial cement replacement in concrete. In: Proceedings of the 13th international conference on environmental science and technology (CEST) conference (2013)
Mustaqim, M.N.M., Song, K.I., Cho, G.C., Zainab, M.: Long-wavelength elastic wave propagation across naturally fractured rock masses. Rock Mech. Rock Eng. **47**(2), 561–573 (2014)
Motamedi, M., Song, K.I., Hashim, R.: Prediction of unconfined compressive strength of pulverized fuel ash–cement–sand mixture. Mater. Struct. **48**, 1061–1073 (2015)
Ogedengbe, K., Oke, A.O.: Pollution impact of cement production air, soil and water in a production location in Nigeria. J. Sci. Technol. **31**(2) (2011)
Rhett, H.: An introduction to electrical resistivity in geophysics. Am. J. Phys. **69**, 943 (2001)
Sansalone, M.J., Streett, W.B.: Impact-echo—nondestructive evaluation of concrete and masonry. Bullbrier Press (1997)
Saxena, S.K., Avramidis, A.S., Reddy, K.R.: Dynamic moduli and damping ratio from cemented sands at low strains. Illinois Institute of Technology (1987)
Song, K.I., Cho, G.C.: Bonding state evaluation of tunnel shotcrete applied onto hard rocks using the impact-echo method. NDT E Int. **42**, 487–500 (2009)
Toma, I.O., Taranu, G., Toma, A.M., Budescu, M.: Influence of cement and sand type on the strength characteristics of mortars with various contents of green binder. Procedia Eng. **21**, 196–203 (2011)
Verástegui-Flores, R.D., Di Emidio, G., Bezuijen, A., Vanwalleghem, J., Kerseman, M.: Evaluation of the free–free resonant frequency method to determine stiffness moduli of cement-treated soil. Soils Found. **55**(5), 943–950 (2015)
Warner, J.: Practical handbook of grouting: soil, rock, and structures. Willey and Sons (2004)

Assessing Benefits of Using Geogrids in Pavements Founded on Problematic Soils

Steven Williams, Jason Wright, S. Sonny Kim(✉), Mi G. Chorzepa, and Stephan A. Durham

Civil Engineering, College of Engineering, University of Georgia, Athens, Georgia
{slwl335, jcwright, kims, chorzepa, sdurham}@uga.edu

Abstract. Geogrids are becoming a popular alternative for soil reinforcement in highway pavement construction to achieve improved performance in regions with soft problematic soils or with a reduction in aggregate layer thickness to reduce construction costs. To examine the potential benefits of geogrids for soil improvement, measurement of permanent deformation using triaxial tests is used in practice. However, soil subgrade improvement in a reinforced pavement system is achieved by lateral distribution of vertical stresses at the reinforcing layer, through the tensile properties of the geogrid material. Therefore, it is desirable to conduct large-scale testing to more accurately monitor the behavior of soil when geogrid is present. The current study seeks to verify the behavior of geogrid reinforced pavement systems through large-scale wheel tests performed with problematic subgrade soils found in North Georgia. The large scale specimen was prepared in a 6 feet long $\times$ 6 feet wide $\times$ 2 feet deep metal box and consisted of 12 in. of aggregate base overlying 12 in. of subgrade soil. Pressure sensors were installed near the bottom of the aggregate base layer and near the top and bottom of the subgrade layer to monitor stress distributions within the pavement system. This paper presents preliminary results showing vertical stress variations obtained experimentally in aggregate base and subgrade soils under large-scale simulated traffic tire loading. The development of a bench scale system to complement the large scale loading system and allow for microstructure evolution studies is also described.

1 Introduction

Building a stable subgrade is vital for constructing an effective and long lasting pavement system. In general, subgrade soils deemed unsuitable to support a pavement system are excavated to an appropriate depth and replaced with stable granular fill material. However, during the placement of granular fill, the use of geogrids or other geosynthetics has become a common alternative in parts of the United States. The purpose of using geosynthetics is to improve the soil stability and the life cycle of the pavement foundation while reducing material usage by enhancing the load distribution over the underlying weak subgrade soils. Similar studies have found the rate of rutting can be decreased as much as 50% when geosynthetics are placed at the sub-base/base

W. Steyn et al. (eds.), *Pavement Materials and Associated Geotechnical Aspects of Civil Infrastructures*, Sustainable Civil Infrastructures,
https://doi.org/10.1007/978-3-319-95759-3_10

course interface and the half-height of the base course (Bagshaw et al. 2015). Geosynthetics can also be beneficial in providing separation between the aggregate base course and the subgrade. Due to the sensitivity of the stability and load bearing capacity of the aggregate base course, the prevention of fines migration from the subgrade into the base course is critical in order to reduce the likelihood of pavement failure.

A pavement failure can be classified as a structural failure where the structure as a whole deteriorates. Another failure method, termed a functional failure (Holtz et al. 2008) occurs when a continuous breakdown of the pavement causes discomfort to passengers and renders it unusable. The use of geosynthetics offer cost saving benefits. First, they increase the number of equivalent standard axle-load (ESAL) passes that an unreinforced pavement of the same thickness will sustain. The second benefit is the reduction of aggregate base course thickness. Geosynthetics used to stabilize base course materials are found to be more effective with coarser aggregate particles. In accordance with AASHTO M 147, the materials for a base course layer should not have less than 10 percent material passing the number 200 sieve, liquid limit of 25 or less, plasticity index of 6 or less, and a maximum particle size not exceeding one third of the base course layer thickness. If the base course does not meet these criteria, proper interlocking of the base course layer may not occur, without which the load will not be effectively distributed over the subgrade layer causing potential deformation and failure. In addition, previous research has demonstrated that pavement stabilization is more effective for low bearing strength subgrades. When the California Bearing Ratio approaches a value of 8 and above, the benefit of using geosynthetics decreases (Bagshaw et al. 2015).

The purpose of this study is to quantify the reduction of rutting potential and changes in vertical soil pressure distribution in geogrid-reinforced pavement systems over problematic soils such as highly plastic silts. To achieve this objective, a large-scale testing apparatus shown in Fig. 1 was fabricated and installed at the University of Georgia's STRuctural ENGineering Testing Hub (STRENGTH Laboratory). In addition, a bench scale laboratory rutting testing apparatus is currently under development at the Sustainable Geosystems Systems Laboratory at the Georgia Institute of Technology with the goal of complementing the findings from the large-scale test system with information on microstructure evolution while also adding to the overall database of experimental results showing the effects of geogrids on pavement performance. Further, the information obtained with the bench-scale system will be used to investigate potential correlations between the rutting behavior and soil geotechnical properties including liquid limit, percentage fines, undrained shear strength and resilient modulus to ultimately aid Georgia Department of Transportation (GDOT) in improving design guidelines and incorporating geogrids in pavements. This paper highlights efforts underway in the development of rutting simulator systems at both universities and presents preliminary results from the large scale system.

Fig. 1. Large scale testing apparatus

2 Materials

2.1 Subgrade Soils and Aggregate Base Materials

For this study, bulk subgrade soil samples were obtained from Gordon County located in North Georgia. The soil is classified as a high plasticity silt (MH) per the Unified Soil Classification System (USCS) or A-7-5 according to the American Association of State Highway and Transportation (AASHTO) classification system. Proctor compaction, Atterberg Limits, and California Bearing Ratio (CBR) tests were performed prior to sample construction to characterize the test soil.

To simulate the worst-case scenario expected during the service life of a highway, the subgrade soils were compacted at a target CBR of 2.5 with moisture contents of 23.6%. The soil has a PI of 21.7. The Graded Aggregate Base (GAB) for the unbound aggregate layer was classified as well-graded gravel or A-1-a based on the AASHTO classification with an optimum moisture content of 6.4% for a maximum dry density of 135.4 lb/ft^3. Sieve analysis and soil index properties for the subgrade soil are shown in Tables 1 and 2, respectively. Figures 2 and 3 illustrate the grain size distribution and CBR value as a function of moisture content of the subgrade soil, respectively.

2.2 Geosynthetic

The geogrid used in the study was selected based on the following factors. Previous studies (Bagshaw et al. 2015) have reported the rate of rut formation was reduced while using a biaxial geogrid. The requirement for geogrid aperture size was determined based on the gradation of the base course material to lie between D_{50} and 2 * (D_{85}) of

Table 1. Sieve analysis of gordon county soil

Sieve #	Sieve size (mm)	Mass of sieve (grams)	Mass of sieve + retained soil (grams)	Mass of retained soil (grams)	% Retained on each sieve	Cumulative % retained	% Finer
3/4 in	19.05	685.30	685.40	0.10	0.01	0.01	99.99
1/4 in	6.35	593.00	657.60	64.60	6.66	6.67	93.33
4	4.76	775.60	794.90	19.30	1.99	8.66	91.34
10	2	470.60	573.60	103.00	10.62	19.28	80.72
20	0.841	495.20	606.70	111.50	11.49	30.77	69.23
40	0.42	388.30	448.70	60.40	6.23	37.00	63.00
60	0.25	332.00	369.30	37.30	3.84	40.84	59.16
200	0.074	501.40	559.80	58.40	6.02	46.86	53.14
Pan	0	366.70	389.50	22.80	2.35	49.21	

Table 2. Soil index properties

	Soil tested
	Gordon county
Specific gravity	2.76
USCS classification	MH
Percentage fines (%)	53.1
Plastic limit	41.7
Liquid limit	63.4
Plasticity index	21.7
Max dry density (pcf)	110.0
Optimum water content (%)	15.7
Water content @ CBR = 2.5 (%)	23.6

the aggregate fill as recommended by Holtz et al. (2008). BX 1200 manufactured by Tensar International chosen for this study. Tensar's BX 1200 Geogrid Specifications are as follows:

- Aperture dimension = 1″
- Minimum Rib thickness = 0.05″
- Tensile strength @ 2% Strain = 410 Ib/ft
- Tensile strength @ 5% strain = 810 Ib/ft
- Ultimate tensile strength = 1310 Ib/ft

These values are minimum average roll values determined in accordance with ASTM D4759-02.

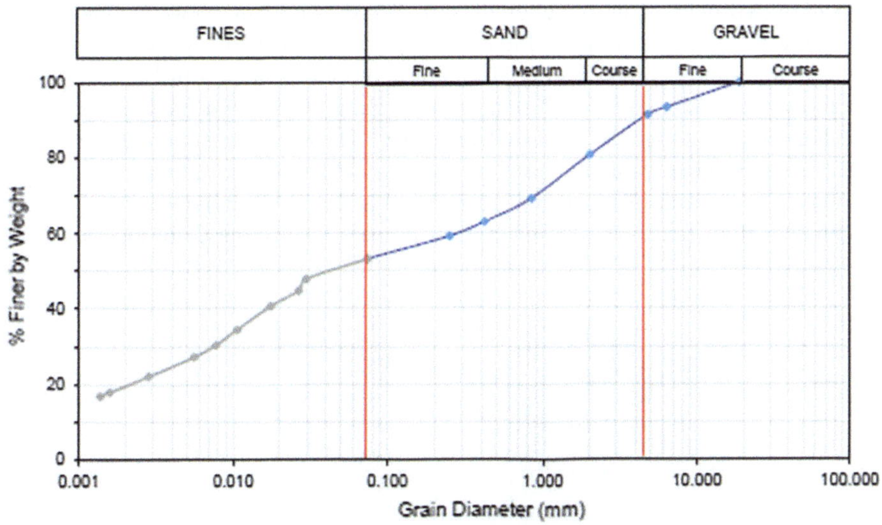

Fig. 2. Grain size analysis chart

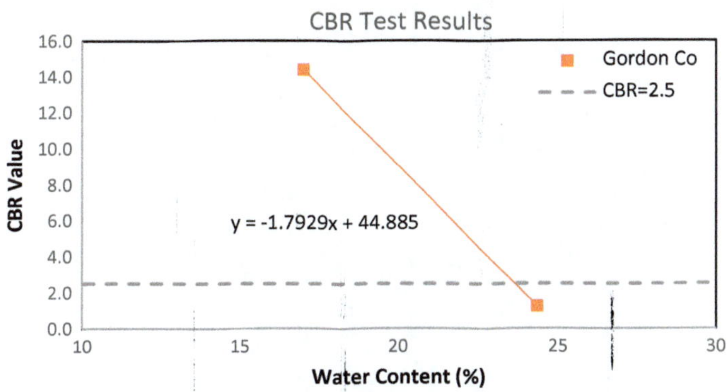

Fig. 3. Variation of CBR values with water content

3 Large-Scale Testing Program

3.1 Large Scale Testing Apparatus

Because of the lateral restraint requirement to utilize the geogrid for pavement foundations, a large-scale, accelerated pavement testing device was developed and utilized for this study. The configuration shown in Fig. 1 allowed for a relatively rapid test preparation and installation as compared to building a comparable highway test section in the field. Each test specimen was prepared in a large metal box measuring 6 feet long × 6 feet wide × 2 feet deep. Specimens were constructed with 12 in. of compacted GAB over 12 in. of subgrade soil. The upper 12 in. along the inside of the walls of the large scale box were lined with a 3-in. thick memory foam to minimize boundary effects during load testing. This configuration is shown in Fig. 4. The large soil samples were fabricated with and without geogrid (Specimens 1 and 2, respectively) between the subgrade and GAB interlayer. The specimens without the use of a geogrid established a valuable baseline against which improvements in pavement performance as a result of incorporating geogrid at the GAB-subgrade interface could be evaluated. Both Specimens 1 and 2 were prepared using the Gordon County soil.

Fig. 4. Memory foam in metal box

3.2 Instrumentation

Each test specimen was constructed with a variety of internal sensors to measure deformation characteristics and performance of the pavement foundation. A total of three pressure cells were installed in each test specimen to continually measure the pressure change with increasing load repetitions. The three pressure cells were embedded directly beneath the wheel load path in order to measure the vertical pressures throughout the different layers. The pressure cell locations are shown in Fig. 5. The first pressure cell was located approximately 1-in. above the GAB-Subgrade interface. The second pressure cell was placed 1-in. below the GAB-Subgrade interface. The final pressure cell was placed at a depth of 20-in. directly below the wheel path. The pressure cells were embedded in fine sand to ensure uniform readings. The geogrid was instrumented with four foil strain gages on the underside of the geosynthetic (facing the subgrade). Two of the strain gages were attached directly in the middle of the specimen between pressure cell locations while the remaining two were located 12 in. from the edge of the testing box. The layout of the strain gages is displayed in Fig. 6. Externally, a motion capture camera was used throughout the test duration to measure permanent deformation and rutting rates with increasing wheel load repetitions.

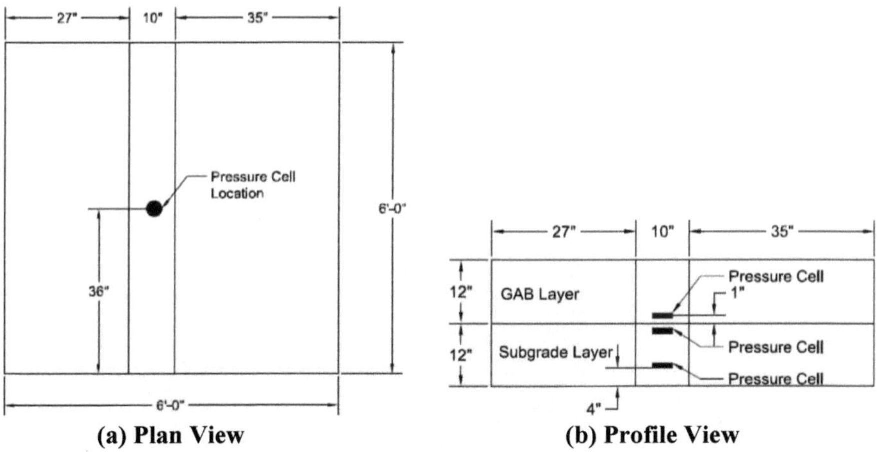

Fig. 5. Pressure cells layout

A load of 2250 lbs was exerted from an MTS actuator onto the large-scale specimen via a truck-size wheel. This load was established based on a finite element analysis model to account for an 8 in. asphalt layer being placed above a 12 in. GAB layer. A standard 9000-pound truck axle load was applied in the model over a 10 in. diameter circular area and approximately 25% of the load was transferred through the asphalt into the GAB layer. Further, it was confirmed through trial testing that the metal box walls do not exert any boundary effects at the stated 2250 lbs wheel testing load.

3.3 Large-Scale Specimen Preparation

The subgrade soils were compacted at a moisture content corresponding to a CBR value of 2.5 while the GAB was compacted to over 95% of the maximum dry density using a plate vibratory compactor. A sand cone test and Dynamic Cone Penetration (DCP) test were conducted in different locations of each layer to ensure proper and uniform compaction. As previously mentioned, three pressure cells and piezometer were installed. Figure 7 illustrates sample preparation stages in progress.

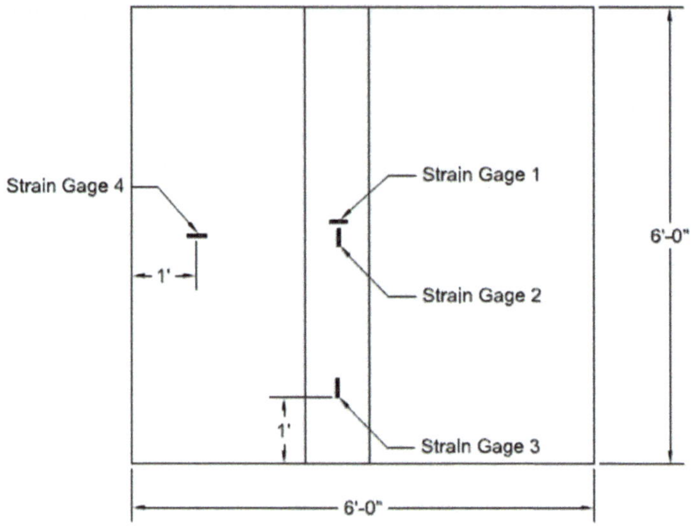

Fig. 6. Strain gage locations

Fig. 7. Large scale testing preparation, **a** Placement of geogrid; **b** Soil compaction; and **c** Installation of pressure cell at the bottom of GAB

Following the construction of each pavement specimen, repeated wheel loading was applied to the specimen at a speed of 1 mph until a 0.4-in. deformation was reached. The vertical pressures and deformation were measured throughout the test duration.

4 Results and Discussion

Figure 8 illustrates the maximum pressures exerted on the pressure cells in the subgrade through 1200 repeated wheel load applications. After approximately 400 passes, the pressure stabilized and remained constant. As expected, the applied force from the wheel distributes onto a larger area as it propagates into the subsurface, and thus the pressure in the sub-layers is reduced for the geogrid specimen.

The vertical pressure measured at all locations showed a reduction of at least 50% when using the geogrid between the subgrade and GAB layers. The purpose of incorporating reinforcing geogrids is to increase shear resistance of the base course thereby reducing the vertical stresses on the subgrade layer. The mechanical concept behind using a geogrid is to produce interlocking of the base course aggregate, thus, increasing its shear strength and bearing capacity and reducing the rutting rate of pavements while distributing the pressure over a larger area into the subgrade layer. Figure 9 illustrates the deformation of the specimens with and without geogrid for a total of 1000 passes. For the preliminary tests, the initial 200 passes were used to ensure equal compaction of the GAB layer for seating purposes. Ongoing research is determining what an appropriate number of passes to achieve seating will be. The data retrieved after 200 passes is presented in this paper. The deformation occurring as a result of 1000 passes decreased by at least 15% when incorporating the geogrid. The test was terminated at 1200 total passes because 0.4 in. of deformation was achieved. However, it is believed that this trend of permanent deformation reduction would continue when using geogrid if more passes were applied.

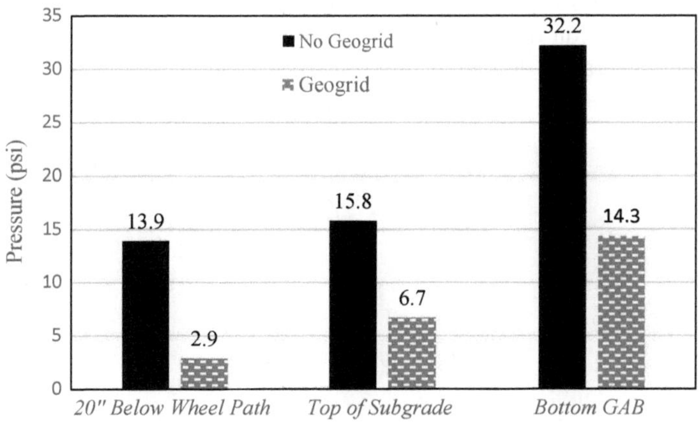

Fig. 8. Pressure reduction

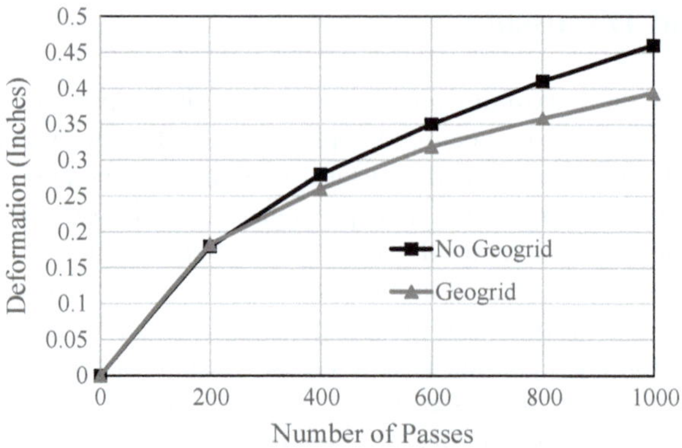

Fig. 9. Permanent deformation after seating load

5 Bench Scale Testing Program

The smaller scale testing system comprises of a 36 in. long, 8 in. wide and 6 in. deep box with a 3 in. diameter wheel at the surface is designed to travel longitudinally over the soil media at a programmed velocity. Vertical deformation along the wheel path and vertical stresses in the pavement interlayers will be continually measured using linear variable displacement transducers and pressure sensors, respectively. Figure 10 shows this system, currently in the final stages of development. While the size of the specimen and magnitude of load will be scaled down for the bench-scale tests, the operating stresses will be the same as the large-scale tests, thereby allowing a direct comparison of the results. The bench scale apparatus also facilitates the study of microstructure evolution during pavement rutting and deformation and permits faster turn-around in generating and analyzing test data.

Fig. 10. Photographs showing small-scale rutting testing system currently under fabrication

6 Summary and Conclusions

The vertical pressure decreases when geogrid is placed between the GAB and subgrade layers. This illustrates the potential effectiveness of the geogrid in regions with poor soil conditions. After the initial 200 passes used for seating purposes, a total deformation of 0.4-in. and 0.46-in. for geogrid and no geogrid specimens, respectively, was measured at the surface during the subsequent 1000 cycles of loading. These preliminary tests confirm that geogrids can be beneficially used in poor soil conditions to reduce the rutting deformation in the subgrade and GAB layers thereby increasing the stability of the roadway system.

These results and associated plans to expand the testing program to include other soft soils encountered in the stage of Georgia as well as tests conducted with the bench-scale rutting apparatus should accelerate the progress towards understanding and improving the feasibility of using geogrid-reinforced pavements where soft-subgrade conditions exist.

Acknowledgement. The work presented in this paper is part of a research project (RP 16-11) sponsored by the Georgia Department of Transportation. The contents of this paper reflect the views of the authors, who are solely responsible for the facts and accuracy of the data, opinions, and conclusions presented herein. The contents may not reflect the views of the funding agency or other individuals.

References

Bagshaw, S.A., Herrington, P.R., Kathirgamanathan, P., Cook-Opus International Consultants LTD, S.R.: Research Report 574 Geosynthetics in basecourse stablisation (Rep. No. 574). NZ Transportation Agency, Wellington, NZ (2015)

Holtz, R.D., Christopher, B.R., Berg, R.R.: Geosynthetic Design and Construction Guidelines (Rep. No. (Bagshaw et al. 2015)). Federal Highway Administration, Washington D.C. (2008)

Tang, X., Abu-Farsakh, M., Hanandeh, S., Chen, Q.: Evaluation of geosynthetics in unpaved roads built over natural soft subgrade using full-scale accelerated pavement testing. Geo-Congress 2014 Technical Papers. (2014). https://doi.org/10.1061/9780784413272.295

Tensar International Corporation.: Product Specification Tensar Biaxial Geogrid (2013)

Warren, K.A., Christopher, B., Howard, I.L.: Geosynthetic Strain Gage Installation Procedures and and alternative strain measurement methods for roadway applications. Geosynth Int **17** (6). Retrieved July 14, 2017. Warren, Christopher, & Howard (2010)

Measuring Specific Heat Capacity of Pavement Materials

Wei Geng[1,2(✉)] and Michael Heitzman[3]

[1] Nanjing Forestry University, 159 Longpan Ave, Nanjing 210037, Jiangsu, China
76400188@qq.com

[2] Nanjing Vocational Institute of Transport Technology, 629 Longmian Ave, Nanjing 211188, Jiangsu, China

[3] National Center for Asphalt Technology, 277 Technology Parkway, Auburn, AL 36830, USA
mah0016@auburn.edu

Abstract. Specific Heat Capacity is an important material parameter that determines the thermal conditions in asphalt pavement materials which influences cracking, rutting, and fatigue. The test standard (ASTM C351) uses a small 25 mm by 50 mm sample container that is not suitable for pavement samples. Research at Arizona State University (ASU) developed a test method that uses standard pavement cylindrical cored samples to measure Specific Heat Capacity. This paper proposes a modified ASU laboratory test procedure for measuring the Specific Heat Capacity of pavement material using a cylindrical core sample, a larger heat exchange chamber, a water agitation device placed above the sample, and multiple thermal couple sensors tied to a data-logger. The heat exchange test is complete in under 30 min. This test method was used for more than 50 samples from four different locations across the United States and the results appear reasonable.

1 Literature Review

Environmental conditions have a significant influence on the performance of pavements, particularly thermal conditions influence the material properties of pavements. Thermal properties, especially thermal conductivity and specific heat capacity, have become increasingly important material parameters to verify or predict the thermal conditions in pavements. Heat Capacity is one of the key thermal properties to describe the transient heat flow of pavement layers [1]. A number of models have been proposed that depend on the thermal properties to estimate the cooling rate [2–4]. Some widely recognized models such as the Enhanced Integrated Climatic Model (EICM) in the Mechanistic Empirical Pavement Design Guide (MEPDG) include thermal properties as part of the process to predict cracking, rutting, faulting and roughness of pavement [5–7]. The research result of the National Cooperative Highway Research Program (NCHRP) project 1-37A [5, 7] (http://www.trb.org/mepdg/) has directly recommended the thermal conductivity and the heat capacity as material input parameters. The problem of Urban Heat Island (UHI) is a research focus in recent years. Some research

W. Steyn et al. (eds.), *Pavement Materials and Associated Geotechnical Aspects of Civil Infrastructures*, Sustainable Civil Infrastructures,
https://doi.org/10.1007/978-3-319-95759-3_11

showed that the pavement material type played a noteworthy role in regards to the UHI [8]. Thermal conductivity and thermal capacity are important parameters that affect the thermal conditions in pavements [9].

2 ASU Cylindrical Specimen Method

The old test method for heat capacity [ASTM C351-92b(1999)] uses a 750 mL Dewar Flask and a 25 mm diameter by 50 mm length sample capsule for measuring specific heat capacity. These test requirements are not suitable for pavement samples because the small sample size would not be representative of asphalt and concrete composite mixtures.

In 2010, J. D. Carlsonand R. Bhardwaj from Arizona State University developed a test method to measure thermal conductivity using cylindrical core samples [10]. In 2011, this method was improved by Morris [11].

In the 2010 test method, all the cylindrical cores were commonly drilled from roads to test the standard mechanical properties and typically have a diameter of 10 cm (4 in.) and a height of 15–20 cm (6–8 in). They also were easily fabricated in the lab.

This size of specimens are useful not just in the thermal conductivity test, but also the heat capacity test. It's quite suitable to test the heat capacity for the asphalt or concrete samples.

3 NCAT Specific Heat Capacity Test

3.1 Summary of Revised Test Configuration

The revised test configuration developed by National Center for Asphalt Technology (NCAT) is based on the ASU Cylindrical Specimen Method, using the same specimens for thermal conductivity test. NCAT is intended to cultivate the test according to the character of pavement specimen. These revisions were made in preparation for efficient testing of more than 50 specimens. Improvements of the test include:

(1) Use the same cores described for the thermal conductivity test. It is easy to extract cores from in-place pavements and also easy to fabricate specimens in the lab. We could use the cores to do both tests without limited specimen preparation. This size of specimen represents the composite nature of asphalt and concrete mixtures.
(2) Use a larger environmental heat exchange chamber (suggested 12-in. deep, 6-in. dia) that can hold approximately equal mass of water and specimen.
(3) Use a stirring system to circulated the water that is inserted above the core specimen during the test. The system has an adjustable speed of rotation and a low heat conductivity plastic stir.
(4) Use a plastic rod the same height of the container to hold the water bath temperature probes. The plastic rod isolates the temperature sensors from the wall of the chamber and the specimen.
(5) Use multiple thermal couples on the specimen and in the water and a data logger during the test and heat exchange is complete in under 30 min.

3.2 Test Preparation

(1) Specimen preparation: Starting with 4-in. diameter drilled field cores, remove the top end of the core (approximately ¾-in. of the surface) and cut to a 5-in. length. If core length is insufficient to obtain a 5-in. length, then cut two cores to 2.5-in. length and bind together with asphalt cement. Drill a ½-in. diameter holed through the center. The ½-in. hole in the center must be carefully drilled to maintain a center location. Asphalt specimens were frozen prior to drilling the hole to improve drill efficiency. This reduced temperature increase from the drill bit that could melt asphalt binder instead of cut through aggregate. The core must be completely air dried and brought to room temperature before proceeding with the test. Weight and record the dry and clean sample as Ms, measured to the nearest 0.01 kg. (The same specimen can be used for heat capacity and thermal conductivity tests).

Prepare approximately a 24-in. length of fishing line with knots at both ends to form two tough circles for finger. Place the fishing line through the hole in the specimen. This line is used to lift and transfer the sample from the oven to the container.

Place two rubber retainers around the sample to fix thermocouples F4 and F5 at the middle height of the sample and make sure the tips of thermocouples make firm contact with the surface of the sample. Make a small hook in the tips of F6 and F7 and put them carefully into the center hole of the sample. Make sure F6 is 1-in. from the bottom of the hole and F7 is 1-in. from the top and the hooks make the tips of thermocouples touch the inner surface of the hole. Place all the four wires together and bend them to be flat at the top of the sample and put the sample with thermal couples together into the oven at the temperature of 50 °C for at least 12 h (Fig. 1).

Fig. 1. Prepared specimen

(2) Water chamber preparation: Cut a 1/4-inch wide and 1-in. deep notch in the top of the container so that the wire leads for thermal couples F1, F2, F3, F4, F5, F6 and F7 could be placed together and get out from the container without any influence of cover lid. Cut a similar notch in the lid and make sure to match the location of the notch in the container. Drill a ½-in. hole in the lid at the center of the opposite half circle with the notch for the plastic stir to pass through. Put a small porous plastic rack (suggest 2.5-in. dia and 1.5-in. height) in the bottom of the container to keep the sample off the bottom of the container and allow easy water circulation to promote release heat.

Fasten three temperature sensors to a plastic rod to measure the temperature of water at different heights. The lowest sensor (F1) is 10 mm above the bottom of the container, the second sensor (F2) is at the mid-depth of the water bath during the test, and the top sensor (F3) is 20 mm below the water surface during a test. Place the water bath temperature probe in the container and arrange the three wires carefully through the notch in the container and fix the stick along the inner side of the container. Make a small bend in the tips of the three thermal couples so the tips do not touch the inner container surface.

Make an insulating sheet (1-in. Styrofoam) at the same diameter of the outer shell of the container and put it under the container to prevent heat loss through the bottom. Use one or more clean plastic containers that can store water for a day of testing and keep at the room temperature at least 12 h before testing. Use a trial specimen to determine and mark the water line in the inner surface of the container before immersing the sample. Once the sample is placed into the water there should be the least air void remaining between the top of the water and bottom of the lid. No water should overflow during the test (Fig. 2). Mark the shaft of the stir rod so that when the stir paddle is in the marked position, the stir paddle could rotate and stir the water as deep as possible without touching the sample or any wires.

3.3 Testing

(1) Weigh the mass of the storage container with water, pour water into the test chamber to the marked water line, weigh the mass of the storage container with remaining water once again, and record the weight of test water as Mw which is the difference in mass of the storage water container measure to the nearest 0.01 kg. Monitor the test water temperature in the test chamber for 3 min then measure a stable water temperature as the average of F1 and F2, and record the temperature as Tw measured to a nearest 0.1 °C.

(2) Record the average temperature of F4, F5, F6 and F7 as Ts to a nearest 0.1 °C while the sample is in the oven. Quickly and carefully transfer the sample from oven to the container using the fishing line to minimize any heat loss during the transfer. Make sure the sample is stable on the base plastic rack at the center position of container and the sample does not touch the inner surface. Then remove the fishing line from the container so that it will not wrap the stir hook. Arrange the wires of F4–F7 carefully through the notch in the test container and make sure the stir paddle will not be influenced by the wires.

Fig. 2. Inner view of water chamber

(3) Place the lid through the hook of stir paddle, match the notches and cover the container, Lower the plastic stir paddle to the marked position. (The mark should allow the pad to be as deep as possible in the water above the specimen and temperature sensor wire leads.) Seal the seam between the lid and container by black electrical tape, turn on the stir equipment to stir the water at a proper speed of rotation so that the heat could equally transfer to the water.
(4) Wrap around the container with a sheet of bubble-wrap and sheet of sponge foam to minimize any heat loss (Fig. 3).
(5) Record the average temperature of F1–F7 as Tm (0.1 °C) when they attain an equilibrium state. This takes approximately 30 min.

The complete assembly is shown in Fig. 4.

3.4 Calculation

$C_{p,s} = \frac{M_w C_{p,w} \Delta T_w}{M_s \Delta T_s}$ Where,

$C_{p,s}$ Specific Heat Capacity of the sample, KJ/kg °C
$C_{p,w}$ Specific Heat Capacity of the water, KJ/kg °C, typical number = 4.2
M_s Mass of the sample kg
M_w Mass of the Water kg

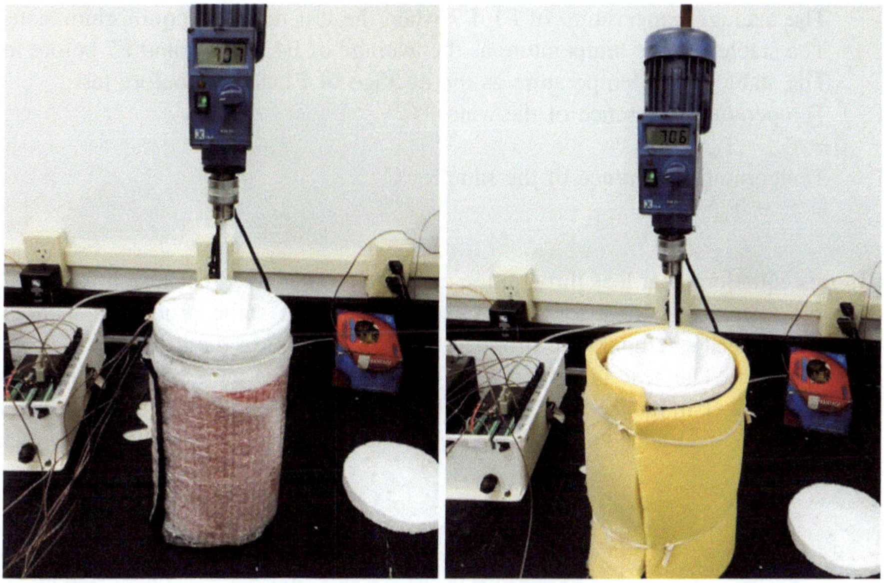

Fig. 3. Heat keeper in the test process

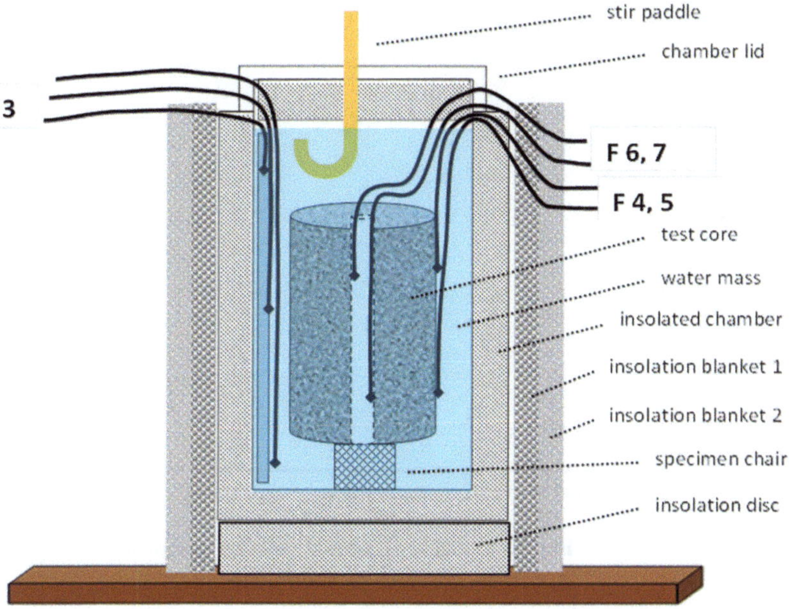

Fig. 4. New diagram of test assembly

T_m The average temperature of F1–F7 when the test reach an equilibrium state.
T_s The stable sample temperature as the average of F4, F5, F6 and F7 before test.
T_w The stable water temperature as the average of F1 and F2 before test.
ΔT_w Temperature difference of the water °C
$= T_m - T_w$
ΔT_s Temperature difference of the sample °C
$= T_s - T_m$

In this equation, heat loss through the container is neglected because it is minimal. Before the formal test, heat loss was checked using the same procedure without a specimen. The water temperature dropped from 101–96 °F in the 4 h (2.7 °C drop in 4 h. See Fig. 5. The test is designed to finish within 30 min, so the temperature difference of water influenced by the heat loss is at most 0.3 °C during the test time.

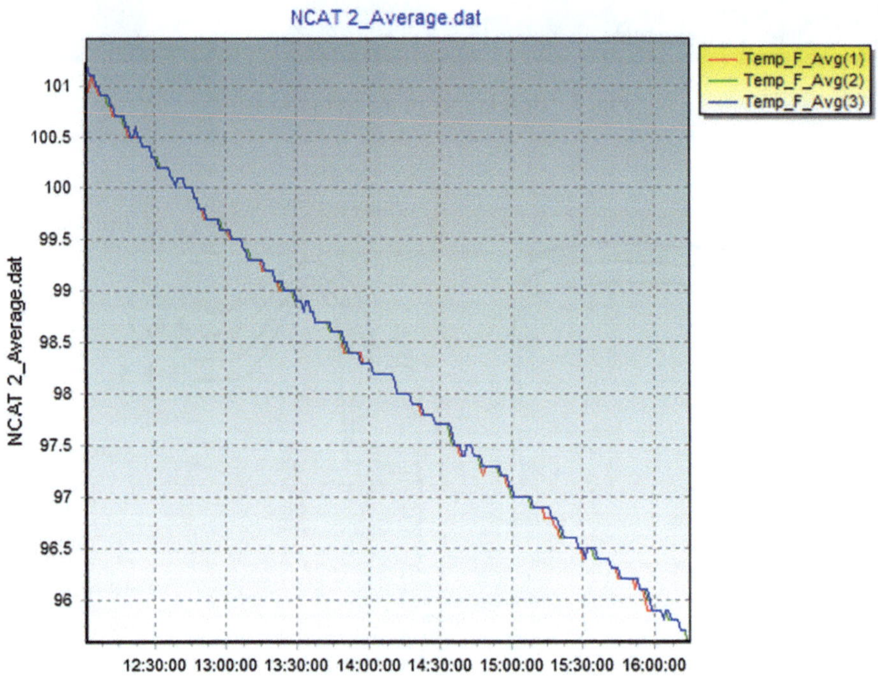

Fig. 5. Container heat loss without specimen

3.5 Test Result

Take the ISU Sample 9 for example. All the test numbers are in Table 1 and the time series curve in Fig. 6. The temperature changes of all 7 sensors are given in the table. F1 and F2 in water are at the same temperature before the beginning of the test. F3–F7 on the sample in the oven are at a similar temperature around 50 °C. When the specimen is transferred into the water, F1–F2 gradually arise and F3–F7 begin to drop. As the water is continually circulated by stirring, the heat gradually transfers from the specimen to the water. Until 18 min later, F1–F7 reach an uniform temperature and the test is terminated.

Table 1. Test number of ISU sample 9

Time (min)	The numbers of thermal sensors						
	T1	T2	T3	T4	T5	T6	T7
0	73.47	73.47	72.42	121.9	122.2	121.3	121.8
1	73.46	73.56	72.8	110	111.7	109.3	110.1
2	73.67	74.75	75.7	84.3	87	81.2	83.1
3	76.93	77.71	77.31	79.32	80.1	77.57	78.59
4	77.97	79.2	78.37	79.56	80.7	78.59	79.13
5	78.78	79.71	79.06	79.94	80.6	79.24	79.66
6	79.34	80	79.55	80.2	80.7	79.68	80
7	79.71	80.2	79.86	80.4	80.7	79.98	80.3
8	80	80.4	80.1	80.5	80.8	80.2	80.5
9	80.2	80.5	80.3	80.6	80.8	80.3	80.5
10	80.3	80.5	80.5	80.6	80.8	80.5	80.6
11	80.5	80.6	80.5	80.6	80.8	80.5	80.7
12	80.5	80.7	80.5	80.7	80.8	80.5	80.7
13	80.6	80.7	80.6	80.7	80.8	80.6	80.7
14	80.6	80.6	80.6	80.7	80.8	80.6	80.7
15	80.6	80.6	80.6	80.7	80.8	80.6	80.7
16	80.6	80.6	80.6	80.7	80.8	80.6	80.7
17	80.7	80.7	80.7	80.7	80.8	80.7	80.7
18	80.7	80.7	80.7	80.7	80.8	80.7	80.7

4 Further Development

In the heat capacity test, all the test results are much close to each other, it still need to figure out the result difference correlating to pavement material properties. In addition, how to reduce the heat loss during the sample transfer from oven to water need further study.

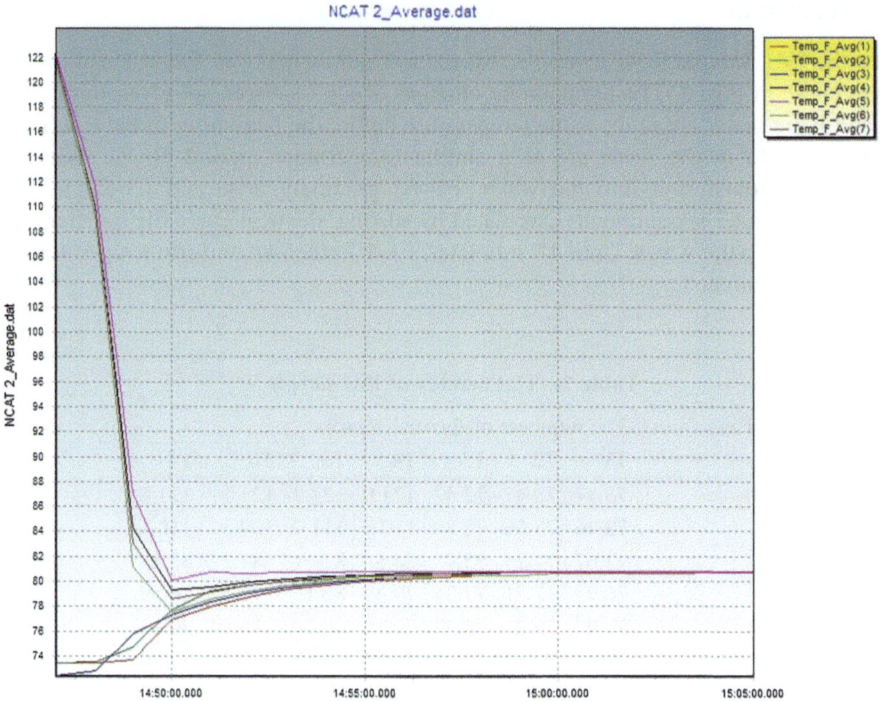

Fig. 6. F1–F7 data curve

References

1. Bruce A.: Chadbourn, Consideration of hot mix asphalt thermal properties during compaction. Quality Management of Hot Mix Asphalt, pp. 127–129 (1996)
2. Corlew, J.S., Dickson, P.F.: Methods for Calculating Temperature Profiles of Hot-Mix Asphalt Concrete as Related to the Construction of Asphalt Pavements. In: Proceedings: Association of Asphalt Paving Technologists Technical Sessions, vol. 37, pp. 101–140. Asphalt Paving Technology (1968)
3. Jordan, P.G., Thomas, M.E.: Prediction of Cooling Curves for Hot-Mix Paving Materials by a Computer Program. Transport and Road Research Laboratory Report 729 (1976)
4. Tegeler, P.A., Dempsey, B.J.: A Method of Predicting Compaction Time for Hot-Mix Bituminous Concrete. In: Proceedings: Association of Asphalt Paving Technologists Technical Sessions, vol. 42, pp. 499–523. Asphalt Paving Technology (1973)
5. NCHRP.: Development of the 2002 Guide for the Design of New and Rehabilitated Pavement Structures: Phase II. National Cooperative Highway Research Program, Project 1-37A, (2004), website (http://www.trb.org/mepdg/)
6. NCHRP report 602; "Calibration and Validation of the Enhanced Integrated Climatic Model for Pavement Design"
7. NCHRP report; Guide for Mechanistic-Empirical Design of New and Rehabilitated Pavement Structures
8. Golden, J.S., Kaloush, K.E.: Mesoscale and micro scale evaluation of surface pavement impacts on the urban heat island effects. Int. J. Pavement Eng. **7**(1), 37–52 (2006)

9. Mouris, M.S.: Using cool pavements as a mitigation strategy to fight urban heat island—a review of the actual developments. Renewable and Sustainable Energy Reviews, pp. 224–240 (2013)
10. Carlson, J.D., Bhardwaj, R.: Determining thermal conductivity of paving materials using cylindrical sample geometry. J. Mater. Civil Eng. © ASCE **22**, 186–195 (2010)
11. Morris, D.: Development of Enhanced Cylindrical Specimen Thermal Conductivity Testing Procedure, Graduate Thesis (2011)

Author Index

W. Steyn et al. (eds.), *Pavement Materials and Associated Geotechnical Aspects of Civil Infrastructures*, Sustainable Civil Infrastructures,
https://doi.org/10.1007/978-3-319-95759-3

MIX
Papier aus verantwortungsvollen Quellen
Paper from responsible sources
FSC® C105338

If you have any concerns about our products,
you can contact us on
ProductSafety@springernature.com

In case Publisher is established outside the EU,
the EU authorized representative is:
Springer Nature Customer Service Center GmbH
Europaplatz 3, 69115 Heidelberg, Germany

Printed by Libri Plureos GmbH
in Hamburg, Germany